MIND SET!

MIND SET!

Reset Your Thinking and See the Future

JOHN NAISBITT

Collins

An Imprint of HarperCollinsPublishers

HarperCollins books may be purchased for educational, business, or sales promotional use. For information, please write: Special Markets Department, HarperCollins Publishers, 10 East 53rd Street, New York, NY 10022.

FIRST EDITION

Designed by Elliott Beard

Library of Congress Cataloging-in-Publication Data has been applied for.

ISBN-13: 978-0-06-113688-7
ISBN-10: 0-06-113688-3

06 07 08 09 10   ID/RRD   10 9 8 7 6 5 4 3 2 1

For Doris

IT IS NOT OUR FEET THAT MOVE US ALONG—IT IS OUR MINDS.
Ancient Chinese proverb

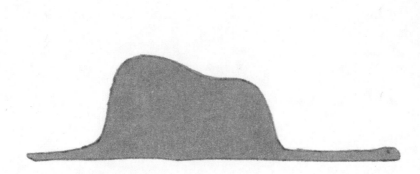

CONTENTS

PART II: PICTURES OF THE FUTURE

The Elephant in the Boa Constrictor

It was a quiet winter evening. I had settled down by the fireplace with my grandson Eli to read him Antoine de Saint-Exupéry's classic, *The Little Prince.*

We were just starting, reading the first page where the six-year-old boy, strongly impressed by a book about the jungle, makes a drawing. To him the picture was clear: It is a boa constrictor digesting an elephant. How surprised he was when the grown-ups did not see what he had drawn—all they saw was a hat! He had to make a second drawing, so they would be able to understand. Only when the boy met the Little Prince did he find someone who looked at his first drawing with the same eyes, seeing it was an elephant in a boa. I had found my metaphor.

Saint-Exupéry's drawings visualize how much our mind can limit what we see. Once we remove the constraints, we will see what is there: an elephant being digested by a boa constrictor.

INTRODUCTION

I grew up on a sugar beet farm in southern Utah in a Mormon community called Glenwood, where most of the no more than 200 men, women, and children were my aunts, uncles, or cousins. As the mountains of Utah enclosed our valley, Mormon rules destined our lives. Life was preordained, and my assignment would be to become a missionary. For two years I would go wherever my church would send me—you don't say no when God is calling you.

I was a little boy when first doubts rose. An ear infection had become chronic and pain had been punishing me with no relief, until my uncle Arnold broke a Mormon rule. Even though smoking was strictly forbidden, he was a smoker. He blew the smoke from his cigarette into my inflamed ears and each time, over more than a year, achieved what no other treatment had accomplished. It eased the pain, and it was entirely outside my mindset. One of my lasting memories of those very early days was the times three or four of my big farmer uncles came to bless me, to heal me, through the "laying on of hands." But never with the comforting result I got from my uncle Arnold.

I started to wonder about the world around me; I experienced a growing desire to explore. What else might there be in this world that my immature mind knew nothing about? I questioned what in my world should have been a given.

Curiosity was the first step, and soon I was sure I wanted to get out of Utah. When I was 17, I saw my chance. I joined the Marines, and, as they say, saw the world, even beyond what they had in mind. The door into this new world truly opened when for the first time I began to read books. It was a world where everything seemed possible.

Since the days I left Utah, the world has been like those books I opened; turning page after page, each page teaching me something new.

Now, looking back these many years later, I can see that my curious mind carried me through many adventures: through being a marine, through my "radical period" as student president of the University of Utah when I fought for free speech and civil rights, my studies at Harvard and Cornell, my time as editor of the Great Books Foundation working for Robert Hutchins and Mortimer Adler, and my three-year cycles in the corporate world in which I left whenever my learning curve began to flatten.

My years in politics instructed me in the art of the possible, from running for Congress at the age of 25 to my time in Washington, where President Kennedy installed me as assistant commissioner of education. After his assassination in November 1963, I stayed on in Washington, becoming assistant to John Gardner when he was secretary of the Department of Health, Education, and Welfare, and eventually working for the White House doing special projects for President Lyndon Johnson.

The 1960s was perhaps the most turbulent decade in U.S. history since the Civil War. Our cities and flags were burning and corporate America was scared. Civil rights and antiwar movements drew millions of people into the streets, and students were occupying hundreds of college campuses.

In July 1964 President Johnson signed the Civil Rights Act to "elimi-

nate the last vestiges of injustice in America." A great bundle of legislation followed. In 1965 I was asked to find out what the impact of all of
his Great Society legislation would have on the country. A fascinating
obligation, but it seemed impossible to find out what was going on at
the time, let alone what the impact of the legislation on America's future
would be. In spite of this compelling consideration, I could no longer
bear Johnson's obsession about winning the war in Vietnam at any cost
("I will not be the first president to tuck tail and run"). I left the White
House for an offer from IBM and became an assistant to Tom Watson,
the chairman.

In April 1967 Detroit erupted in race riots. That summer Black
Power advocates called for an armed revolution, and riots swept across
the country. President Johnson rushed 4,700 paratroopers into Detroit,
which had been nearly paralyzed by race riots. Mayor Cavanaugh said,
"This is like Berlin 1945." The president ordered new training for riot
control of all National Guard units. The turbulence was fuelled the next
year by the assassinations of Martin Luther King, Jr., and Bobby Kennedy. Around that time, the shift to the use of the word *black* instead
of *Negro* began. America's cities were burning, and companies did not
know what was going on and what they should do. This instability fed
my hunt for a mechanism to monitor the transition the country was
going through. How could I really see where America was headed?

One day I bought a copy of the *Seattle Times* at an out-of-town news-
stand in Chicago. The headline revealed that the local school board had
voted a new reform package. My eyes went over the different local head-
lines of newspapers that the kiosk sold from all parts of the country. I
suddenly realized that by reading all those local newspapers every day,
I could recognize new patterns of change in the country. I could really
figure out what was going on in the United States. I had found the key—
my epiphany.

My entrepreneurial genes kicked in. I left IBM and Tom Watson, a job some of my colleagues would have cheerfully killed for, and with my last paycheck started my own company, Urban Research Corporation.

Before long my growing staff and I were monitoring and cataloguing 160 local daily newspapers. We focused on local events on the notion that the sum of what was happening locally was what was happening in the country. It was a multitude of occurrences that I saw through my own experiences and my way of thinking at the time. The summary of what we learned and what we made of it was published in a weekly report: *Urban Crisis Monitor*. Soon many companies became subscribers, and I began to give speeches about what I thought was going on in America.

In the following decade, I kept working with major corporations and learned more about American society through our content analysis method. Out of the highly localized database, I watched the general outlines of a new society slowly emerging and got a sense of the direction in which America was restructuring.

Although for me the changes were clearly under way, some of the new directions I saw seemed unlikely at the time. I risked displeasing the experts, who could have argued that describing the world in terms of shifting categories would be too simplistic. But I thought it was worth the risk. I felt that in a world where events and ideas were analyzed to the point of lifelessness, where complexity grew by quantum leaps, where the information din was so high that one had to shriek to be heard above it, people were hungry for structure. With a simple framework, we could begin to make sense of the world. And we could change the framework, as the world itself changed. All of this finally resulted in my writing *Megatrends*.

Megatrends sold 9 million copies. *Reinventing the Corporation* and several other books followed, including *Megatrends 2000*. I gave speeches and lectured on all continents. Wherever I spoke in the world, journalists, politicians, and members of the corporate world began to ask, "What will be the next megatrend?" My audiences wanted to know where we will be in the year 2010 or 2030. And with an increasing insistence, they asked, "How do you know?" and "How do you do what you do?"

MINDSETS

To a little boy with a hammer everything looks like a nail.

It was an afternoon I spent with my friend Toni Ofner when he would not stop digging into the question of how I did what I do. He would not accept my answer that the information I collected, the many newspapers I read, the ideas and thoughts I exchanged with many people in different countries and cultures around the world made me see where the world was heading.

"Well, then," he said, "if, as you say, the future is embedded in the present, and if I carefully watch what is going on in the world, I should get to the same results you do. But I don't. So what makes the difference?"

I paused for a moment; I had not really looked at it that way. But as I thought about it, it became clear to me that the difference was not in what I learned but in how I thought about it. "I guess, Toni," I heard myself saying, "it's my mindset." I had been aware that over the years I had developed certain rules to discipline my mind and filter information. I match and measure information against my own experience, using my values and mindsets. And so does everybody else. "What you say," Toni said, "sounds like rain falling on different ground."

He was right; it's true that the mindsets are the ground on which the rain (information) falls and the plants that sprout differ depending on the mindsets we have and thus lead to different conclusions. It is *how we receive information*. That is the key.

We went on with our discussion, and I began to think about what the most powerful and important mindsets were that had helped me or held me back. Later I started to play with the idea of basing a book on mindsets and how they guide me to get pictures about the future.

Judgments in almost every area are driven by mindsets, from world affairs to personal relationships. If a wife's mindset is that she has a philandering husband, she receives all information as fitting into this picture. It defines what she hears beyond words and her reaction to it. If a wife's mindset is that she has a loving, faithful husband, she receives the same information with a different meaning. That's at a micro level.

At a macro level there are those whose mindset is that the world is in a period of a "clash of civilizations," and they see everything within this framework. Others, including me, see the world through a mindset of a long period of economic determinism, a "look to the economics."

Of course we all have mindsets that are inculcated as we go through life: politicians are all a bunch of crooks; outsourcing to India is stealing American jobs; cats are the cleanest pets you can have; global warming is a threat to the sustainability of humankind.

But in this book, I do not write about mindsets that are the result of acculturation or driven by social coercion. I focus on mindsets that are deliberately developed for a purpose. You can create mindsets that can instruct and organize you in your personal life and field of endeavor. Thus, this book provides not only the framework and the perspective of the first half of this century but also the fundamental attitudes that are necessary to anticipate the future, to receive the future.

Part I: Mindsets

Mindsets work like fixed stars in our heads. Holding on to them, our mind drifting like a ship in an ocean of information, finds orientation. They keep it on course and guide it safely to its destination.

The mindsets I describe in Part I of this book helped me to adjust and correct my thinking; they removed the constraints I had and help me get the most out of the information I collect. Without those mindsets, it would not have been possible for me to get to the results of *Megatrends* and *Megatrends 2000*. *How* I approached things, how my mind processes information and experience, was the key.

To each person, one or another of the 11 mindsets I will introduce to you might appear to be the most important and helpful. My premier mindset is "Understand how powerful it is not to have to be right." It is a great release in any field of business and private life, indispensable in any endeavor where you venture out. It is the mindset that will enable you to dare to say or try whatever you are working on, no matter how unlikely it seems at the time. It was the liberating mindset behind the success of *Megatrends* and the books that followed. It is a mindset that supports creative imagination.

My second important guide is the mindset "Don't get so far ahead of the parade that people don't know you're in it." It sounds so obvious, so easy, but as examples later in the book will show, it is not easy to keep the margin just right. In every field of life, in business, in leadership, or in politics, it is the mindset that helps you not to extrapolate so far ahead that people do not relate to what you do or say.

In daily work, all mindsets work together, each meshing with the others, bringing clarity in a confusing world, enabling you to see not only a hat but also the elephant in the boa. You will experience them as cognitive tools that can transform your life and business.

Part II: Pictures of the Future

We all like to have a picture of the future to get clarity in a confusing world. We often ask, "What will be?" Part II of this book will give you an answer. It deals with the major trends ahead of us and with their impact on our business and life.

How will commerce, trade, production, and services be affected in a world where literacy and the novel are in decline and visual communication is in ascendancy? Are nation-states and their economic indices any longer important for your field of endeavor and well-being? How do we anticipate where Europe is headed when the rhetoric paints one picture while the score of the game shows a quite different one? Will China, the omnipresent job-eater, be the dragon that devours us or the dragon we ride? Are we about to see a Next Big Thing as the hype in the media promotes?

My pictures of the future are not speculation or a reach into the unknown. They are based on an analysis of the present, under the guidance of the mindsets described in Part I. You will be able to recapitulate why I arrive at certain conclusions and how I keep focused on the whole and don't get lost in details or things that will only be blips on the big picture screen of the world. My pictures of the future will give you the frame, which you can apply to your own fields of interest and endeavors in order to profit from the future.

PART I

MINDSETS

Part one of this book will lay the ground to reset your thinking. In applying the 11 Mindsets, you will sometimes experience the need to discipline your mind and sometimes feel you just need to let your mind flow.

While many things change, most things remain constant

IN A 24/7 MEDIA WORLD, THE HYPE IS CHANGE

On August 8, 2006, Amazon.com listed 56,170 book titles under *change*, 11,195 titles under *business change,* and 2,404 titles under *global change.* An uncountable number of newspapers, magazines, and 24-hour news channels leave not one stone unturned, promoting the idea that everything is changing. Now, who in the world can keep up with this? No one can.

Don't bother.

Think about it: Most businesses stay in a steady state, day in and day out, year in and year out. Yes, products and markets have been altered, mostly for the better, and the tools we use have changed. But despite the avalanche of business books, business practices—the basics of buying and selling, of making a profit as a necessary condition of survival—have remained much the same during my 40 years of involvement.

Whether cell phones can display television and calls are made via the Internet, your bathtub filled by taking off your clothes, or your refrigerator opened by a rumble in your stomach, these are just other ways of doing what we do—easier, faster, farther, more, and longer—and not the substance of our lives. We go to school, get married, and have kids and send them to school, which, God knows, does not change despite the chanting for school reform. Home, family, and work are the great constants.

Life on a sugar beet farm has not changed too much since my boyhood. As ever, the seasons determine the rhythm of life, although modern equipment has eased sowing and harvesting. Most of the farmers still raise chickens and keep farm animals, only the horses in my time used for transport and work are more often kept for leisure and delight. In the ups and downs of life my parents were trying to make their living, educate their children as best they could, all in the frame of their potential and with the tools of their time.

At the beginning of their 11-volume *Story of Civilization*, Will and Ariel Durant say:

> Civilization is a stream with banks. The stream is sometimes filled with blood from people killing, stealing, shouting and doing things historians usually record, while on the banks, unnoticed, people build homes, make love, raise their predictable children, sing songs, write poetry and even whittle statues.
>
> The story of civilization is the story of what happened on the banks.

THE DIFFERENCE BETWEEN WHAT AND HOW

Have you ever made a list of what changes or is likely to change? Hold on—not a list of *how* we do things, but *what* we do.

Returning to farming, what has changed is how agriculture is practiced. Advancement is dependent on how flexible farmers adapt their skills to new technologies and changing consumer behavior. But farmers stay what they are, farmers, although there are differences in how they farm. Some find their niche, adapting to changing demands in the market, like Chino's, whose delicate organic vegetables and fruits are flown to celebrity cooks like Wolfgang Puck in Los Angeles. Others, for various reasons, have not made it and have given up farming.

Most change is not in what we do, but how we do it. Within all the hype, the more we are able to differentiate between constants and change, the more effectively we will be able to react to new markets and profit from change.

Sports are a good example.

The rules of team sports remain fairly constant—with only very small changes from time to time. The changes we do make often come from a change in the way players play their game. A well-known big change was the popularization of the modern forward pass in football by Knute Rockne in the 1920s. The goal was still to get a touchdown, but how players could get to the goal line changed.

Occasionally an individual player's style will change a game.

On the night of December 30, 1936, a crowd of more than 17,500 turned out at the old Madison Square Garden in New York City, to see Long Island University, the nation's number-one basketball team with a 43-game winning streak, oppose Stanford, the defending Pacific Coast Conference champion. Stanford ended LIU's winning streak with a 45–31 victory, but something more important happened.

The crowd in fact had mostly come to see Hank Luisetti, Stanford's

6 foot 2 inch, 185-pound sophomore. He was the only player known for shooting the ball with one hand while he hung in the air, in defiance of basketball style. Everyone else was shooting the old style: two-handed set shots or hook shots. The huge publicity celebrating Luisetti's shooting style did not change that the goal was putting the ball into the basket, but it forever changed how the game was played. But not without stubborn resistance. The establishment felt it was not the right thing to do. "That's not basketball," Nat Holman, the fabled City College of New York coach, said at the time. "If my boys ever shot one-handed, I'd quit coaching."

Luisetti was voted college player of the year in 1937 and 1938. He finished second to George Mikan in the Associated Press's poll to select the best player of the first half of the twentieth century.

Hank Luisetti died on December 17, 2002, living plenty long enough to see his style perfected and embellished by the likes of Earl Monroe, Julius Erving, and, of course, Michael Jordan.

A change in techniques often opens a door to a wider potential. American track-and-field athlete Dick Fosbury literally developed a jump into a new era. Instead of leaping facing the bar and swinging first one leg and then the other over the bar, Fosbury turned just as he leaped, flinging his body backward over the bar with his back arched and his legs following, landing on his shoulders. Fosbury, a high school student in Medford, Oregon, started to high-jump using the straddle method he learned from his coaches, but his performance was rather mediocre until he started to look for his own way. He did not think about the traditional school of jumping, not about form. "I don't even think about the high jump," he said. "It's positive thinking. I just let it happen."

At the 1968 Olympic Games in Mexico, his head-first technique of jumping captivated the audience when he cleared every height up to 7 feet 3¼ inches. With 7 feet 4¼ inches on his third attempt, Fosbury set a new Olympic record and won the gold medal.

The sudden progress in high jumping was not based only on this new technique. The new technique was made possible by a change in the landing area, from sandpit to large rubber beds, allowing a head-first jumper a landing without killing himself. Although the sport itself had not changed, Fosbury's innovative style, the Fosbury flop, became the universal method.

THE HIGH MORTALITY OF CHANGE

So often we hear and read that "the only certainty is change." There is an upside and a downside to this. The upside is on the side of the change management consultants. The downside is that people are driven into hysteria about the omnipresence of change. In the late 1990s e-commerce was to change everything. Sell all your stock in retail chains; forget bricks-and-mortar.

Fashion, one would think, is all about change. Most of fashion, I would say, is a parade of fads. But there are many constants in the fashion world, and I don't mean the little black dress. A constant of more than 150 years is jeans, which were "Levi's" for about 100 years until the options increased. Now every designer has his or her jeans version coming and going.

A constant joy in men's fashion is its constancy. Don't try to introduce something new. This market, unlike women's fashions, hardly ever changes, and if it does only slightly. In my experience over the years, just about the only change has been the width of men's ties—every 20 years. More things are like men's fashions than women's fashions.

As the caprices of women's fashions change, so do the fragrances enveloping them. Their life span often is as elusive as their fragrance. Of 100 new fragrances, 95 fully vaporize. But more keep being produced because the markup is so high that if you are one of the 5 survivors, it's great. The big change in fashion is that it has become increasingly interwoven with art and architecture, turning it into a whole new experience of visual art. You will read about the many faces of the new visual art world in Chapter 1.

In November 2005, Coca-Cola announced that it was pulling the plug on Vanilla Coke, joining the long list of new products that die an early death. Of the 30,000 new customer products launched each year, more than 90 percent expire. "We are doing just fine with what we have" seems to be the response of consumers. The importance of continuity was highlighted in *Built to Last*, the 1994 book by Jim Collins and Jerry Porras, in which they debunked the "myth of change." "A visionary company almost religiously preserves its core ideology, changing it seldom, if ever," they wrote. "Core values in a visionary company form a rock-solid foundation and do not drift with the trends and fashions of the day."

In business, as in sports, new techniques sometimes introduce a new constancy. "Process management" took off in the United States in the 1980s, triggered by the threat of the Japanese taking over the world, and with its star methodology, Six Sigma, pioneered by Bill Smith at Motorola, became the new universal truth for quality improvement. Now its great champion, GE's Jack Welch, has retired, and it is beginning to run out of gas, as it is now being seen by many as a hindrance to creativity. But it has been in play for 25 years.

EVERYTHING NEW UNDER THE SUN?

Long gone are the days when King Solomon (3000 BC) wrote his famous words: "That which has come to be, that is what will come to be; and that which has been done, that is what will be done; and so there is nothing new under the sun."

Now everything is said to be new under the sun.

Early in 2005 *Newsweek* started running a widely placed advertisement with an extraordinary quote by its gifted international editor, Fareed Zakaria: "The 21st century will be the century of change. More things will change in more places in the next 10 years than in the previous 100. Most countries aren't ready for this dizzying ride—certainly not the United States of America."

No wonder reading statements like this make us focus on the future, desperately seeking the horizon for the next sign of change, each little cloud a possible sign of a thunderstorm. Fareed Zakaria might be able to look beyond the horizon, seeing more than any one of us. But surely no quantitative studies were made: Adding up all the changes during the previous 100 years is daunting enough, but it is not possible to now list the number of things that "will change in more places in the next 10 years." It is an attention-getting statement in the fog of speculation. To me it is impressionistic and not helpful. Better to make your list of what you think will change in the next 10 years and what you think is likely to remain constant. The good news is that "the only constancy is change" is both ubiquitous and ridiculous. My bottom line is that "the only certainty in business is change" is just not true.

THE DNA OF CHANGE SPIRALS AROUND
THE PILLARS OF CONSTANCY.

Change is the nourishment of the media. Its appetite is fed by competition and 24/7 reporting: the relevance of an event is mostly the event itself, its quality and significance depending on the supply and easily gliding into triviality and irrelevance.

A common purpose of the 11 Mindsets in this book is to not get lost in the nonessential, but instead to focus on the things that have and will have the strongest influence on our lives.

Most of us are not hunting for news and change, but for orientation into the future, for clarity in a confusing world. Not quantity but quality is decisive. Whatever information is assailing you, distinguish between real and apparent change, basic shifts and fads, remembering that in the history of the world, most things remain constant.

Differentiate between

> Basics and embellishment
> Rules and techniques
> Trends and fads
> Breakthroughs and refinements

The future is embedded in the present

JOHNNY-LOOK-IN-THE-AIR

As he trudged along to school
It was always Johnny's rule
To be looking at the sky
And the clouds that floated high

But what just before him lay,
In his way,
Johnny never thought about

Came a little dog one day
Johnny's eyes were still astray
Down they fell, with such a thump
Dog and Johnny in a lump

Mark Twain translated the old German rhyme, which for me is a metaphor of how we, focusing too far out into the future, can stumble over what is right in front of us. Mark Twain probably came across "Johnny-Look-in-the-Sky" in the 1890s, when he was traveling to Europe and spent 20 months in Vienna.

Many trendsetting ideas and theories were born in Europe during this period, reaching far into the future. Architects, poets, and painters were rebelling against stolid traditions. Otto Wagner, Adolf Loos, and Josef Hoffmann founded the Jugendstil and revolutionist architecture; Gustav Klimt and his associates began setting "rivers of modern art aflame"; Sigmund Freud searched the unconscious mind and became the father of psychoanalysis. Many seeds of today's world were laid and started to foliate.

World wars interrupted, but could not stop the path once taken. For many, things would not turn for the better. The wounds the war struck in Germany would be extended by reparation payments, as predicted by the great economist John Maynard Keynes. He was one of the people who carefully studied the present and therefore was prealerted about events that eventually really happened.

Keynes participated in the negotiations of the peace treaty of Versailles after World War I, although he strongly disagreed with the high war reparations imposed on Germany. In his book *The Economic Consequences of the Peace* (1919), he predicted that the reparations required of Germany would lead to catastrophe; comprehensible today but rejected and denied for various reasons at the time.

In 1945 Friedrich Hayek, the Austrian economist and early champion of the idea of a free market, wrote *The Use of Knowledge in Society*, in which he predicted the collapse of Communism. He did so for a solid reason: Successful central governing would need to know the decisions made by the market before they were made, which of course is impossible, a subject about which you will read more in Chapter 4, "Europe."

Keynes and Hayek never drifted into the unknown, searching the sky for undetected stars; they searched the territory within range. Their works are examples of how only an objective and unbiased study of the present can reveal the future. The hostility and denial they faced from politicians and established colleagues show how much certain mindsets can becloud and cheat one's view.

The insights of three great Austrian-born economists, Joseph Schumpeter, Friedrich Hayek, and Peter Drucker, could not prevent Europe from becoming a master of denial. Promises and predictions of European Union politicians ascend into the sky like balloons until they burst when colliding with reality. They continue their empty promises because of the widespread mindset of having to be right, which is nowhere more cultivated than in politics.

Keep that in mind when you focus on what is really embedded in the present. Search for solid ground on which to start your considerations. In all my books I have worked that way. But be careful, the future being embedded in the present certainly does not mean extrapolating everything into the future. It is not saying that the future will be little more than an extension of things as they are. It is saying that we find the seeds of the future on the ground, and not in the width of the sky.

In doing so, watch out for some traps, search the ground carefully, and take advantage of the help available, as the following pages will show.

YOU CAN'T SEE THE FOREST FOR THE TREES

A few years ago I gave a speech in Panama City in connection with the hand-back of the Panama Canal from the United States to Panama.

As a warm-up to the ceremony, during which former President Jimmy Carter did the honors of officially returning the canal, Panama's minister of foreign affairs, who hosted the meeting for the visiting foreign ministers to whom I gave my speech, invited my wife Doris and me to join a helicopter excursion over the canal. We are scared of helicopter flights, but to avoid being wimps, we could do nothing but agree.

Of course we had read how difficult the building of the canal had been, how many lives it had cost, but the understanding of the dimension of the project came with the view from the height of the helicopter. An impenetrable rainforest spread for miles, a sight of splendor and foreboding, a denseness that made the jungle under us look like a lush green carpet, through which the blue man-made band of the canal was drawn.

What a different sight a few days later when we were taken on a hike through the rainforest we had seen from above: a richness of roots, trees, palms, plants, and bushes, of different greens, leaves, and flowers, accompanied by the chirping, whistling, and twittering of birds and the chattering of monkeys jumping in the treetops, an opulence of details changing with every turn of the path. It was another world, with different perspectives and insights, but also proof that you lose the view of the forest if you are entertained by the trees.

The distinction between details and the big picture cannot always be experienced in such an exuberant way. But the idea stays the same: You cannot see the forest for the trees. If you want to spot precedent-setting events in the world, you have to monitor it from a distance. If you lose distance, fads can easily block your view. Fads themselves are embedded in trends and are a manifestation of trends. Shifts in trends do not occur very often, but the fads that the trends have embedded inspire change all the time—thus the word *faddish*.

One of the grand shifts I wrote about in 1982 in *Megatrends* was the shift from an industrial society to an information society. Embedded in this shift was a trend that was most likely to follow, a "body awareness" trend. In the agricultural society, we worked our bodies hard doing our daily chores. Similarly, during our industrial period we did physical work on the job (the short history of the United States: farmer, laborer, clerk). As we moved into an information era, the demand on physical efforts shrank dramatically. Now, in today's information society, most of us make very few demands on our bodies during our workday. Among other things, this shift has meant that we have become predominately sedentary workers, more and more concerned about what we do with our bodies and what we *put into* our bodies, an extraordinary era of body awareness. The shift to being sedentary workers was a consequence of the larger shift to an information society and will not change.

Given our sedentary condition, we will continue to deal with our bodies, but *how* we deal with our bodies is much less important and is subject to fads. These fads are part of the implementation of the trend, the changing manifestations of the trend. For a long time, running was very big for keeping in shape. Swimming had its day; now many think Lance Armstrong is the way to go, and personal trainers and fitness boutiques are proliferating. How we keep in shape, the different things we try, comes and goes, and can be very faddish, but the necessity to make an effort to keep our bodies toned stays. Accompanying this physical consideration is concern about what we eat, our diet.

Diets and dieting, of course, are outrageously faddish: Mayo Clinic, Atkins, vegetarian, watermelon, Weight Watchers, South Beach, Beverly Hills, Scarsdale, low-fat, high-fat, only carbs, no-carbs, lots of fruit, no-fruit. Despite the recent attention paid to the Atkins diet, low-carbohydrate diets have been around for a long, long time. In 1825, Jean Anthelme Brillat-Savarin, a French lawyer, in a book called *The Physiology of Taste* that was a sensation in Europe, warned that some of his readers would object to his "diet":

"Oh Heavens!" all you readers of both sexes will cry out, "Oh Heavens above! But what a wretch the Professor is! Here in a single word he forbids us everything we must love, those little white rolls from Limet, and Achard's cakes, and those cookies. He doesn't even leave us potatoes or macaroni! Who would have thought this of a lover of good food who seemed so pleasant?"

"What's this I hear?" I exclaim, putting on my severest face, which I do perhaps once a year. "Very well then; eat! Get fat! Become ugly and thick, and asthmatic, and finally die in your own melted grease."

While not as colorful as Brillat-Savarin's, the diet pitches keep coming. There seems to be no end to opportunities to contrive shortcuts to keeping the weight off. But the fads are transient ways we deal with the consequences of sitting on our butts all day. You can see fads as manifestations of a major trend; you can see trends coming from the major shifts in a society. Several examples in Chapter 1, "A Visual Culture Is Taking Over the World," will help you to reconstruct how to arrive at such conclusions.

LOOK BEHIND THE CURTAIN

Most of us played hide-and-seek as children. One of my hiding places was behind a curtain. Some of the major shifts are not much harder to spot than a boy behind a curtain—but only if you look.

Another of the shifts I wrote about in *Megatrends* was that union membership would continue to decline and that the labor movement was dead. Why did I say that?

In the early 1980s, one quarter of the American workforce was organized in unions. Hidden behind the curtain of equality of treatment was a profound mismatch between the philosophy of the labor unions ("treat everyone exactly the same") and the emerging *individualism*. All this was happening in the context of a big shift away from manufacturing employment where most of the union membership was located (but not away from manufacturing itself, which was being increasingly automated); at the same time there was a growing affluence among workers, who sought less organized protection. It made labor unions look like dinosaurs waiting for the Mesozoic Era to return. It did not. The political ground had shifted beneath them. Like the dinosaurs, the unions would have had to reinvent themselves, reconceptualize their role, if they were to survive. They did not.

From the time I wrote about the decline of U.S. labor unions until today, the percentage of the workforce that was unionized went from 25 percent to only 7.8 percent in the private sector and is still falling. The just-announced split-off of the service workers unions and the Teamsters from the AFL-CIO plunges labor unions into further decline.

Basic change is the result of a *confluence* of forces, rarely because of just one force (especially when it is against the received wisdom). Always ask yourself if there are enough different forces at work pushing in the same direction, before you make a judgment. No conclusion in Part II of this book is made on single evidence. If you don't find enough evidence, you'd better look behind the curtain again.

A PLANETARY RESEARCH DEPARTMENT

The great source of knowledge to lift the curtain about the future is newspapers—operating as a planetary research department, reporting what's happening on the ground. Newspapers, as many people have said, are the "first draft of history." Nothing else comes close. That's why newspapers are a primary source for those who study the past and those who are interested in the future. History books often list contemporary newspapers as primary source material, but we rarely think of today's newspapers as primary source material. Look at newspapers as if you were reading them 100 years from now. Of course what we look for in newspapers are not opinions but the recording of events and scores of the games. How we find the score of the game, and what we have to watch out for, is the subject of the next Mindset.

Newspapers are forced choice in a closed system. Let me explain. Newspapers are great monitors of social change because, simply stated, the *news hole*—the space available for news stories in a newspaper—is a closed system. For economic reasons, the amount of space devoted to news in a newspaper does not change significantly over time, so when something new is introduced, something must be omitted or reduced. You cannot add unless you subtract. This is the principle of forced choice in a closed system. The managing editor of the *New York Times*, John Geddes, said that the size of the news hole at the *Times* "hasn't changed in eight years or so," underscoring that "it is a zero-sum game." The online bloggers have no such "space budget." They just go on forever without regard to length or content.

There is a similar forced-choice situation in societies. In a sense it is a changing share of market that competing societal concerns command. Societies are like human beings: A person can keep only so many problems and concerns in his or her head or heart at any one time, just as a society can deal with only a certain number of considerations at any one time. If new, compelling problems or concerns are introduced, some existing ones are diminished or given up. All this is reflected in the collective news hole that becomes a mechanical representation of society sorting out its priorities.

While it is not only what you read but also how you read it, some newspapers are much better than others. For me, the best chroniclers of change in the world today are the *New York Times, Financial Times, Wall Street Journal*, the *Economist* (which calls itself a newspaper), and *USA Today* (which is very strong on creating box scores for many aspects of society).

But do not count on even the above newspapers placing the score of the game on the pages of their publications that reflect the importance of the scores in the long run. For example, in late March of 2006, Ford Motor Company announced that it would lay off 30,000 workers. It was the big front-page story in almost all papers, even though the layoffs would be carried out during a three-year period and even though it was not an unexpected development from declining Ford. On the same day the U.S. Commerce Department announced that the U.S. economy had added 280,000 new jobs in February (a short month), a development of much greater impact for the long-term future. I found the new-jobs story relegated toward the back to page 10 or 26 in all the newspapers I saw; reported, but buried. Put the weight on the importance of the occurrence, not on its placement reflecting concerns of the moment.

In the stream of time, the future is always with us.

The directions and turns the world will take are embedded in the past and in the present. We often recognize them retrospectively, but our purpose is to anticipate what lies ahead. To do that successfully, we need to keep a distance and a clear eye.

Newspapers are our great collaborators. They are not only the first draft of history but the first to give us a glimpse of the future, because what we are doing now will determine the future. Newspapers are the basic source of the information and geographic range. Newspapers provide stories and facts about politics, culture, social affairs, events, trends, and fads. But they also carry personal opinions, political propaganda, irrelevant details, and temporary blips that are not likely to be signals about the future. While it is crucial to be well instructed, it is not the amount of information we collect but how consciously we receive it. In a verifying and selective process, we can find the components that bring together the pictures of the future.

Focus on the score
of the game

SPORTS VS. POLITICS 1-0

How do you read newspapers? I read newspapers back to front, sports first—not so much because sports are a celebration of human achievement and will be more important in an increasingly technology-driven world (a matter I deal with in Mindset #11), but also because I love sports. I have an abiding interest in the subject. In addition, not insignificantly, my bias is that the reporting on sports is the most reliable—and less so as you move toward front-page news. If I read that the Boston Red Sox beat the New York Yankees 7–3, I can be almost 100 percent sure that that actually occurred.

Over the years at the other end of the paper, often on page 1 of the *Financial Times*, Chancellor Gerhard Schröeder blustered about an economically strong Germany. World affairs, of course, are not as transparent as sports, but they have statistics, too. Inside the paper, I found the scores of the game: Unemployment in Germany rose by 44,000 the previous month, taking the unemployment rate to 10.4 percent. But the government announced that Germany's growth in 2005 would be around

1.5 percent, little enough. But look at the score of the game: Germany's economy was not growing at all—zero—and the previous quarter was negative.

One of the first things that Angela Merkel did in January 2006, after she had been chancellor of Germany for two months, was to order the German economics ministry to break the habit of producing over-optimistic growth forecasts. A good beginning.

A LACK OF REALITY

Back in 2000, the heads of the governments of the European Union met in Lisbon, where they pledged to make Europe "the most competitive and dynamic knowledge-based economy in the world by the year 2010." Europe obviously was challenging the United States for global economic supremacy. Each year, the leaders of Europe have since renewed that pledge, now known as the Lisbon Agenda, balloons of promises and predictions rising high in the sky.

What is the reality? What is the score of the game? In every single year since the announcement in 2000, Europe has lost economic ground against the United States. Indeed, for more than a *decade*, Europe has each year lost ground to the United States. No matter how compelling the proclamations of the ambitious goals of the European Union, the steps needed to make them a reality are much more important. If you want to know how the European Union and its member countries are doing and where it and they are headed, you have to constantly check the scores of the game. Is employment rising or falling? How much growth is being achieved? Are any economic reforms being carried out? Few jobs can be created by governments; it is entrepreneurs who create new jobs. How nourishing is the environment for starting a company? Is productivity increasing, or not?

Whatever you read about Europe, keep asking those questions. The balloons only fly till they burst.

A BACKPACK FULL OF BURDENS

While Europe in a way wears its social struggle on its sleeve, General Motors, the leftover icon of the fading American industrial might, hides its struggles in a backpack. In January 2006, General Motors (worldwide revenue of $192.6 billion) posted its largest annual loss in more than a decade—$10.6 billion—laying bare the problems it is facing: rising labor costs, fierce competition from Asia, and falling sales at home. GM is in a horrible bind. The carmaker's backpack is filled with a $1,600-per-vehicle handicap in so-called legacy costs, mostly health and pension benefits.

Worst of all, GM reached a watershed in its four-decade decline in market share. Toyota is about to replace it as the world's biggest automaker; GM is now cash flow–negative. For the first time since the early 1990s, it actually consumes more cash than it brings in making cars. As summarized in *BusinessWeek*:

> Normally a company in such straits contracts until it reaches equilibrium. What would a healthy GM look like? It might have five fewer assembly plants, building around 4 million vehicles a year in North America instead of 5.1 million. That would slash U.S. market share to around 20 percent, but factories would hum with real demand, stoked less by rebate giveaways and cheapo rental-car sales. Workers would have a cost-competitive health-care plan but would fall back on government unemployment benefits when hard times demanded layoffs. Profitable auto sales and finance operations would fuel a richer research budget, tightly focused on four or five divisions instead of eight.

But for GM, shrinkage is not much of an option. Because of its union agreements, the automaker can't close plants or lay off workers without paying a stiff penalty, no matter how far its sales or profits fall. It must run plants at 80 percent capacity, minimum, whether they make money or not. Even if it halts its assembly lines, GM must pay laid-off workers and foot their extraordinarily generous health-care and pension costs. Unless GM scores major givebacks from the union, those costs are fixed, at least until the next round of contract talks in two years.

GM's payroll pumps $8.7 billion a year into its assembly workers' pockets. Directly or indirectly, it supports nearly 900,000 jobs—everyone from auto-parts workers to advertising writers, car sales-people, and office-supply vendors. When GM shut down for 54 days during a 1998 labor action, it knocked a full percentage point off the U.S. economic growth rate that quarter. So what's bad for General Motors is still, undeniably, bad for America.

More than 100 years ago, Henry Ford's Piquette Plant (at the corner of Piquette Avenue and Beaubien Street in Detroit) started the modern U.S. automotive industry. Bill Ford now faces the possibility of presiding over its demise. In January 2006 Ford's CEO made this extraordinary statement: "From now on, our products will be designed and built to satisfy the customer, not just to fill a factory." *Imagine!*

In business, politics, or private life, the gap between words and facts widens when personal pride is involved. Often it's not the promises made but the problems hidden. In the fight for performance, the power of *having* to be right often takes over. Don't be misled; check the score of the game.

That is what strategic advisors do. Companies pay millions of dollars to advisors to essentially tell them the score of their game. McKinsey & Company and other consultants analyze the numbers—the facts and statistics—of the company, analyze the efficiency of every unit and worker, analyze the market the company operates in, and then provides fact-based advice. They are paid so much because most of the time, they are right. You can do the same thing by analyzing the scores of the games in the world around you.

THE FLAG OF EXCELLENCE

The score of the game might include the number of Nobel Prizes awarded—at least in science, which are less political than those in literature or peace.

Before World War II, Europe won 109 Nobel Prizes in science, while only 13 went to the United States. Since 1969, however, that has been turned around: Europe has received 90 prizes, and the United States, 171—a huge shift in global scientific leadership. I will come back to this in Chapter 4 about Europe.

Several years ago when lecturing in Tokyo, I was talking about how immigration was constantly replenishing America's talent pool and later noted that it wasn't by chance that the United States had more than 300 Nobel Prize winners and Japan had only four. A guy in the second row said, "Yeah, but most of those Nobel winners moved to the U.S. from other countries." "Thanks," I answered, "for making my point."

COLD? WARM? COLD?

When I was a boy growing up in Utah in the late 1930s and the 1940s, the great fear was "the coming ice age." The rhetoric then and again in the 1970s was about "global cooling"!

In 1974 American Pulitzer-winning columnist George Will wrote,

"Some climatologists believe that the average temperature in the North Hemisphere may decline two or three degrees by the end of the century. If that climate change occurs, there will be mega-deaths and social up-heaval because grain production in the high latitudes (Canada, northern regions of China and the Soviet Union) will decrease." Well, the end of the century has come and gone, and now it is supposed to be getting warmer.

Also in the 1970s—30 years ago—a group of leading climatologists, meeting in Bonn, warned that "the facts of the present climate change are such that most optimistic experts would assign near-certainty to major crop failures within a decade [because of global cooling]. If national and international policies do not take these near-certain failures into account, they will result in mass deaths by starvation and probably in anarchy and violence that could exact a still more terrible toll."

The best-selling book of this period was *The Cooling: Has the Next Ice Age Begun? Can We Survive It?* published in 1975 by Prentice-Hall. Its author, Lowell Ponte, who also lectured widely on the subject, said that some of the world's most conservative scientists have warned of an ice age beginning in the near future. "Global cooling," he said, "presents humankind with the most important social, political and adaptive challenges we have had to deal with for 10,000 years. Your stake in the decisions we make concerning global cooling is of ultimate importance: the survival of ourselves, our children, our species."

In 1975, Nigel Calder, editor of England's *New Scientist*, said that "the threat of a new ice age must now stand alongside nuclear war as a likely source of wholesale death and misery for mankind."

Well, you get the idea: the same rhetoric, the same maniacal urgency.

Some of the same scientists who were hysterical about global cooling are now hysterical about global warming.

In the fall of 2005, British Airways announced it would give passengers a role in helping to fix the world's climate by allowing them to pay a few dollars extra on their ticket, the money to be used to offset the carbon emissions from their trip. It turns out that fewer than 1 in 200 passengers were willing to participate. This doesn't seem to square with the answers people give to opinion polls about climate change. Does it, however, square with what economists call "revealed preferences," the idea that talk is cheap but that people's decisions where their money is at stake is the best guide to people's true beliefs?

How will proclamations of environmental doom be read in the decades ahead? On January 28, 2006, former President Bill Clinton said in an enthusiastically applauded speech at the World Economic Forum in Davos: "We must make a serious global effort to develop a clean energy future to avoid the onset of another ice age."

Then on February 4, 2006, the U.S. Minerals Management Energy Report noted: "Polar bear deaths are a strong sign of global warming's effects. Researchers from the U.S. Minerals Management Service are reporting increasing numbers of polar bears that are drowning, evidence of the effects of global warming on the polar ice caps."

THE DILEMMA OF THE NONEXPERT

Who and what are we to believe? Where to begin? Read a significant number of the 963 books on global warming listed on Amazon.com, and, to keep a balance, read from the 1,054 books (July 2006) on global cooling and the coming ice age?

Amid all the conflicting claims and advice, it is sometimes difficult to know whom to believe about the environment. I cannot stop my life to spend the next 2 or 10 years to become an expert on the environment or sustainability. Nor can any of us. I can only use my experience and best judgment.

The debate is exacerbated by the superior tone of those who are so sure about global warming. Global warming has become a religion, and those who don't buy into its gloom and doom scenarios are infidels who must be banished from any public forum.

I believe the environment must be protected and that regulation is often necessary. No matter who is right about the environment and sustainability, I support attending to the environment because the remedies are so attractive. I want clean air and clean water for everybody. I vote for nature. But exaggerating problems without any real idea of the score of the game distorts society's priorities and makes it hard for citizens and leaders to make the best decisions.

WHOSE ROOSTER CROWS THE LOUDEST?

Environmentalists routinely exaggerate problems so as to alarm people and get support for their agenda. In his 1980 book, *The Sinking Ark*, Norman Myers, an ecologist at Oxford University, estimated that one species would become extinct every second day. In 1999, Paul Van Develder of the *Seattle Times* wrote: "We know that twenty years later, zoologists and botanists estimate that the various activities of Homo sapiens have accelerated the rate of extinction to a mind-numbing, spirit-crushing (are you ready for this?) 75 species per day."

That adds up to 27,375 a year. Soon the number was inflated to losing 40,000 species a year. Now, your intuition and common sense tells you that this cannot be true. But that figure appears again and again in the media. It is based not on any evidence but on Dr. Myers's wild assertion in a 1979 article that one million species might be lost between 1975 and 2000, which works out to be 40,000 species a year. The 40,000 number has been endlessly and uncritically repeated in the media. It has a life of its own.

The World Conservation Union, which maintains a record book of endangered species, said in 1992 that the extinction numbers for mammals and birds were very small and that the total extinction rate—

assuming 30 million species—was probably around 2,300 species a year. But no one really knows.

In the meantime, Dr. Myers recently repeated his 40,000-a-year estimate, warning that "we are into the opening stages of a human-caused biotic holocaust." His irresponsible and reckless disregard for scientific integrity is typical of a number of environmentalists and also shows disdain for nature's sustainability powers.

Horror scenarios wear off if repeated too many times without the event occurring. Some danger, on the other hand, is not attractive enough to defend its ranking in our daily bad-news reports.

In the Shade of the Shark

It was her second day on the island. The sun was rising from the gently rushing ocean; birds intoned their early morning songs. She was one of the first on the beach, wandering between the coconut trees enjoying the cool breeze. Death came on silent soles. A little snap as it detached from the tree, a sound she had not noticed before a fraction of a second later the nut crushed her head with a crack.

Breaking news on CNN? Killer coconut strikes again? The brown devil kills tourist? Not at all. Coconut killers don't sell. They don't have terrifying teeth, no fin cutting the water causing panic on the beach, just the rules of ripeness, gravity, and coincidence working. Despite the neglect more people are seriously hurt or die from falling coconuts than from shark attacks, where any incident at any time finds it way into the media. "Shark! Shark!" sells.

According to George Burgess, director of the University of Florida's International Shark Attack File and a noted shark researcher, "Falling coconuts kill 150 people worldwide each year, 15 times the number of fatalities attributable to sharks." In fact, Burgess's latest statistics show that in 2002, 2003, and 2004, the combined figure for worldwide shark attack fatalities was only 14.

Shark attacks are news, falling hairy brown nuts are not. It is not the probability but the theater of dying. And even when the neglected plant of horror finally gets into the limelight, it is in connection with a prize in the shadow: the IgNobel Prize.

IgNobel Prize

"Injuries Due to Falling Coconuts," was the title of an article run by the *Journal of Trauma,* in which Dr. Peter Barss reported on nine injuries in Papua New Guinea, in this case none fatal. Dr. Barss points out that the average palm tree in that part of the world reaches a height of 75 feet, and that a coconut can weigh four pounds or more, meaning that a coconut probably would fall at a velocity of about 50 miles an hour, with a force of about 2,000 pounds. Several victims suffered fractured skulls.

The article received an IgNobel Prize, given annually at Harvard by the editors of the *Annals of Improbable Research* in recognition of research that "cannot or should not be replicated." The award was presented in 2001, although the paper had been published in 1984. Apparently news takes a while to get to Cambridge.

I believe it will take until we have the first "Killer Coconut" movie before their victims will get out of the shade of the shark.

The Circus of Sensations

Bad news often develops a life of its own. Scores change, not very often in sports but frequently in life, especially with disasters when politics of several kinds get involved. Often with catastrophes, early high or low figures don't prove out. In these cases, it is prudent to wait until reliable returns are in. The score was truly tragic for the Asian tsunami of 2004, where the eventual death toll grew to more than 250,000. Quite often, though, the scores fall when opinion, not truth, is the father of the figures.

The first numbers coming from Hurricane Katrina, centered in New

Orleans in 2005, climbed to an estimated 10,000 fatalities, with some reports saying the final toll might be as high as 100,000. We now know that the death toll was about 1,000, terrible enough.

The Chernobyl accident of April 1986 is another case in point. When the nuclear reactor in Ukraine exploded, there were apocalyptic alarms sounded that animal herds in greater Europe would be decimated. Germans and Austrians, among others, were vaccinated against dying from radiation sickness. Some experts forecast that there would be as many as 800,000 additional cancers in humans in the decades ahead. New studies by the United Nations and government agencies now reveal that 56 people have died from causes related to Chernobyl radiation, 47 of whom were plant personnel killed by the blast or the fire that followed. About 4,000 children did develop thyroid cancer, almost all of whom were cured; only nine died.

It is in the nature of human beings to bend information in the direction of desired conclusions.

THE SCORE OF THE GAME CANNOT BE BRIBED.

Keep sports as the model. When a football game ends 27–17, that's the score of the game. The result does not change because of excuses, praise, or explanations from the losing or winning team. In business, companies do not perform better because of the rhetoric of their CEOs. Geopolitics and cultural activities are shaped not by what is said but by what is done or not done. What is happening on the ground is what matters. Political game scores are dealt with in Chapter 3, "China" and Chapter 4, "Europe." Focus on the score of the game you are dealing with, compared with the box scores in baseball and basketball. Their simplicity and reliability is the standard against which to measure the accuracy and relevance of information. Complexity is often used as an instrument of camouflage, simplifying supports transparency. All of this is an aid to understanding the present, which is the first step to understanding the future.

Understanding how powerful it is not to have to be right

A MODEST MAVERICK

It was not a great start for the young man. He had studied physics at the Polytechnikum Zurich, but his dissertation was turned down. "You are quite smart," his professor had told him, "but you have one big failing. You never listen to anybody." Sounds like an appropriate comment about someone who believed that "authority slavery is one of the biggest enemies of truth."

At the beginning of the twentieth century, such an attitude was not exactly a door-opener and the young man was the only one of the four graduates in his class who did not find an academic job, despite sending out stacks of applications. Nevertheless, his self-confidence was unbroken: "Long live insolence! It is my guardian angel in this world." After an intermission as a teacher in a school in Schaffhausen, he finally was hired as a patent examiner in Bern. He was Albert Einstein.

Beginning in 1902, he spent 48 hours a week behind his desk in the patent office convinced that he was born to create something great, in the evening hours, if necessary. In 1905 Einstein wrote a letter to the mathematician Conrad Habicht, his friend from the times in Schaffhausen:

Dear Habicht,

Such a solemn air of silence has descended between us that I almost feel as if I am committing a sacrilege when I break it now with some inconsequential babble. So, what are you up to, you frozen whale, you smoked, dried, canned piece of sole? Why have you still not sent me your dissertation? Don't you know that I am one of the 1.5 fellows who would read it with interest and pleasure, you wretched man? I promise you four papers in return.

The first deals with radiation and the energy properties of light and is very revolutionary, as you will see if you send me your work first. The second paper is a determination of the true size of atoms. The third proves that bodies on the order of magnitude 1/1000 mm, suspended in liquids, must already perform an observable random motion that is produced by thermal motion. The fourth paper is only a rough draft at this point, and is on the electrodynamics of moving bodies, which employs a modification of the theory of space and time.

What the 26-year-old Einstein dismissively called "inconsequential babble" was some of the most astonishing babble a human being has ever uttered. It catapulted physics from the nineteenth into the twentieth century and transported its author to be one of the century's greatest geniuses. How was it possible that one man could shake up the fundamentals of physics? Certainly it was not only his youthful high spirits that made him unique. In the "annus mirabilis," the miracle year, of 1905, as we now call it, he not only presented the theory of relativity but also the second big shift in thinking in quantum mechanics. With these two papers, he had created a new foundation on which his successors would build the house of modern physics.

As if these efforts had not been enough, Einstein topped them at the end of the year. Habicht received another letter about an idea that Einstein called not only captivating but also funny: "I wonder if God is going to laugh because he was jokingly misleading me."

$E = mc^2$ was born.

Einstein illuminated the hidden universe of time and space. And as one of his biographers, Albrecht Fössling, speculated, it was an advantage that "a man of such condition did not bother to get entangled in discussions with leading scientists, but was hanging out with his thoughts in his own diaspora." Albert Einstein took the freedom to imagine, the freedom to make connections otherwise not seen, willing to be surprised by any result. He focused on substance, not ego.

Many have tried to explain the genius of the Bern patent examiner. Harvard intelligence specialist Howard Gardner believed Einstein discovered so many secrets of nature because he never lost his childlike naïveté and called him an "eternal child." Certainly Einstein's big concern was not whether his attempts would be disproved, but to miss a fork in the road. On September 22, 1911, he wrote to William Julius:

Highly Esteemed Colleague:

If these [solar spectral] lines are very fine, then I believe that my theory is refuted by these observations. I would be very pleased if you told me candidly your opinion about this matter. After all, I know very well that my theory rests on a shaky foundation. The road I took might be the wrong one, but it had to be tried out.

On October 27, 1912, he wrote to another acquaintance, the astrophysicist Erwin Freundlich:

My theoretical studies are progressing briskly after indescribably painstaking research, so that the chances are good that the equations for the general dynamics of the gravitation will be set up soon. The beauty of the thing is that one can keep clear of arbitrary assumptions, so that there is nothing to be "patched up"; instead, the whole thing will be either true or false.

At the brink of a ground-shaking discovery, the temptation to fight for being right gains many advocates. But to stand up and overturn established concepts of time and space, energy and matter, leaves no room for a need to be right. Einstein, with all his feelings, fears, and hopes, was in a quest for universal and immutable laws, for what was right, not who was right.

Few of us reach out as far as Einstein, but the need to let go of having to be right is not diminished with more modest goals. What is right, not who is right, stays the standard: in private life, in business, and in politics.

When I write about the future, I can't worry whether my conclusions might be proven wrong at some later date. In the end, I have to make a judgment, and that judgment has to be as well instructed as possible. But only if I don't have to be right can I imagine anything, suggest anything. Nevertheless, some statements seemed quite unlikely at the time I wrote them.

DARE IT

One of the conclusions I came to in *Megatrends* that caused a vigorous reaction was that the high-flying national television networks—NBC, CBS, ABC—were on a long downward slide because their audiences would be siphoned off by the emerging—but then quite primitive—

offerings of cable television. At the time (1982), to think that those net-work entertainment giants would do anything but become more impor-tant was laughable to most people. I remember giving a speech about this at the annual meeting of the Advertising Association of America (right in the lion's den) at the Greenbrier, and the next day the advertising col-umnist for the *New York Times* wrote about my speech, saying, in effect, that "anyone who believes this is some kind of wacko." Well, maybe I would turn out to be wrong, but it seemed to me that general-purpose instrumentalities in America were losing ground everywhere. The great general-purpose magazines *Life*, *Look*, and the *Saturday Evening Post*, with their 10-million circulations, had folded, replaced by hundreds of special-interest magazines. It seemed pretty straightforward to me that ABC, CBS, and NBC would become the *Life*, *Look*, and *Post* in the years ahead, their audiences drawn away by the increasing options of improv-ing cable television. Today, the combined market share of the so-called three major networks hovers at about 11 percent.

DRIVING AGAINST TRAFFIC

Can everyone else be wrong?

For decades the almost universally accepted medical theory was that ulcers were caused by stress, smoking, and alcohol. The idea was that stress led the body to produce excessive stomach acid, which ate through the stomach's lining. A standard treatment was surgery.

Then two Australian physicians, Robin Warren and Barry Marshall, said that ulcers were caused by a then unknown strain of bacteria, a statement as outrageous as driving against traffic on a superhighway in-sisting that everyone but you was going the wrong way.

In 1983, Dr. Marshall began successfully treating ulcer sufferers with antibiotics. Later that year at a disease conference in Belgium, Dr.

Marshall was asked if he thought bacteria caused some stomach ulcers. Dr. Marshall dramatically answered that he believed that bacteria causes *all* stomach ulcers. Are all experts traveling the wrong way? How could that be?

"It was impossible to displace the dogma," he later said. He was booed off the stage at that and other scientific meetings. "Their agenda was to shut me up and get me out of gastroenterology and into general practice in the outback," said the young doctor from Perth. The worldwide medical community stood firm against the impertinent idea of an ulcer bug.

In October 2005, Drs. Warren and Marshall were awarded the Nobel Prize in medicine for discovering the "ulcer bug." It had taken more than a decade for the establishment to give up prescribing antacids and start using antibiotics. It is awesome to even think about calculating the collective cost and pain caused by the medical establishment's determination not to have been proven wrong.

Having to be right shackles your mind.

People are culturally conditioned to have to be right. The parents are right, the teacher is right, the boss is right. Who is right overrules what is right. Couples have huge quarrels about considerations that are forgotten as the struggle for who is right rages on.

Political parties have institutionalized having to be right. How often has a political party welcomed the position of the other side? Imagine if all the energy that goes into trying to prove the other side wrong were channeled into actually thinking about what was best for whatever the dispute is about. Worse, having to be right becomes a barrier to learning and understanding. It keeps you away from growing, for there is no growth without changing, correcting, and questioning yourself.

If you have to be right, you put yourself in a hedged lane, but once you experience the power of not having to be right, you will feel like you are walking across open fields, the perspective wide and your feet free to take any turn.

See the future as a picture puzzle

SEQUENCE IS THE ENEMY OF MAKING CONNECTIONS

Eugene Fubini, IBM's brilliant director of research at the time when I was there, relished starting his scientific seminars with the question "What comes next in this sequence: 4, 14, 23, 34, 42?" No one ever got it right. The answer is 50, the next stop on the Eighth Avenue line of the New York City subway system after 42nd Street. Fubini always caught his audience in a numbers sequence mindset. Sequence originates from the Latin *sequo*, meaning "to follow," and describes the consecutiveness of things similar: in music, repeating a motif in a different pitch; in movies, the meaningful order of scenes; in mathematics, a sequence following a formula. But to explore, we have to make connections between things that seemingly don't fit, are not obviously related, and sometimes seem to contradict common sense or formulas.

PICK THE RIPE FRUITS

Who has ever solved a picture puzzle by first putting the pieces in a straight line?

My history teachers taught history in a sequential way. The past was lined up year by year and country by country. At the end, my class-mates and I could rattle off dates and events, but how events, dates, and countries were connected and feeding on each other, we could not see. We only knew the retrospectively created march of history. Historians and commentators often fall back on a neat series of events, one step following another, to explain the past. But the past, the present, and the future are a thicket of bobs and weaves and lurches and sputters as we stumble through our lives; in a world "that is now so interpenetrated," as Salman Rushdie said, that "to explain a murder in California, you have to understand the history of Kashmir." Only statistics can be extrapo-lated—not the future—and it is well known how they can be made to distort a picture. Making the connections is more intuitive than calcu-lated.

Each breakthrough breaks old mindsets. That's the character of change. Discoveries grow out of something that is already there. Ripe apples had always fallen to the ground, but Isaac Newton saw "the deeper meaning." The Earth has always circled the sun, but Copernicus and Galileo observed the evidence and made the connections.

Geniuses often build on details that many people can spot but can't connect. Einstein was a master at detecting and connecting. During his university years, he laid the groundwork and studied the masters of theoretical physics with "holy eagerness." Soon he had collected an enor-mous overview, from the very large to the very small, from the cosmos to subatomic particles, that he continually renewed by studying the latest scientific reports. He built on this ground, putting in question what his predecessors had established, and pointed out logical contradictions in the structure of physics. But what was known before Einstein? What important step took him beyond everything ever thought?

Hardly any of the formulas Einstein used in his famous work "On the Electrodynamics of Moving Bodies" were really new. Many, in fact, had become common property among experts. Berlin historian Dr. Jürgen Renn, director of the Max Planck Institute, brought it to a point: "Many say Einstein has created the theory of relativity out of nothingness. But in reality he did nothing but pick ripe fruits." This should not diminish Einstein's epochal role. Renn continues: "The question remains, why he chose to pick the fruits he did. Even if Einstein's greatest talent was to be a gifted fruit picker he could not have chosen a better time. Never before had so many ripe apples been hanging on the tree of physical cognition. But they were also hanging so hidden, that only a very sharp eye could track them down."

Indeed, most "ripe fruits" can be seen by many in different fields. But only by making the proper connections can the individual "fruits" result in an understandable picture and begin to make sense.

THE FUTURE IS A PICTURE PUZZLE

The future is a collection of possibilities, directions, events, twists and turns, advances, and surprises. As time passes, everything finds its place and together all pieces form a new picture of the world. In a projection of the future, we have to anticipate where the pieces will go, and the better we understand the connections, the more accurate the picture will be.

Sometimes—in an effort to make randomness my friend—I write down on cards individual events, phenomena, fads, and surprises. I then shuffle the cards, fan them out, shuffle them again, each time noticing if the new random juxtapositions give me some new ideas, some new connections, that I had not otherwise noticed, a new way of arranging facts into an illuminating new pattern.

It makes no sense to put the pieces in a straight line. We have to find the parts that fit with each other, that intertwine and connect. We don't start empty-handed. There are always the constant parts, the things that don't change; those are your base. Then you have to sort out the pieces that are relevant for the subject, for the puzzle you are focusing on, and begin to search out which sides match and complete each other. It might take many trials, but events that seem totally unconnected when lined up in a row begin to show their coherence when they are connected with their matching part. Einstein might be a high reach, but in the simplicity of a genius, he selected and collected the ripe fruits and put them in their place. On whatever level we do what we do, the principles of the 11 Mindsets in this book will help us to sort out, judge, and place the single pieces.

The comment I heard the most when *Megatrends* was taking off went something like this: "I sort of knew about most of those things in your book, but you put all the pieces together for me." I have never heard a description I liked more of what I do than "picking ripe fruits." The question is what to pick and where to put them. The purpose of *Mind Set!* is to help you to do that.

When I started my company, Urban Research Corporation, the puzzle I had to solve was to find out what was really going on in America. As you know from the Introduction, I had my epiphany at the right time. I knew that America is an overwhelmingly bottom-up society and that the pieces of my puzzle would be found in the local newspapers from around the country reporting local events.

Trends and ideas begin in regional cities and local communities such as Tampa, Denver, and Seattle, not in big-city centers like New York or Washington. Monitoring such highly localized data, I saw the emerging patterns of a new society slowly being shaped. Each piece of information, each puzzle piece, then found its place. When I read about the high technology of heart transplants and brain scanners in our hospitals and, separately, about more massage and prayer in those same hospitals, I knew they matched and completed my formula of "high tech/high touch." I saw the interplay, the connection between the introduction of technology and our counterbalancing response. The pieces merged into a picture of how we seem always to learn to balance the material wonders of technology with the spiritual demands of our human nature.

MIX AND MATCH UNTIL YOU SEE THE NEW PICTURE.

If we want to anticipate the future, there are always some puzzle pieces to begin with, the first things that catch our attention. The 11 Mindsets will give you guidance to pick the right fruits and form the picture of the future you're interested in. Look for the score of the game pieces and don't be afraid of making mistakes placing them. Continue to examine the pieces for their connectedness with other pieces, and those fitting together will slowly shape the new picture.

All five pictures of the future in Part II of this book have been built that way. Chapter 1, for example, introduces you to the eight pieces that led me to the conclusion that "A Visual Culture Is Taking Over the World." Chapter 2 focuses on how the pieces in the world's economic puzzle will be shuffled to new ways of measuring economic data and achievements. Chapters 3 and 4 are dedicated to the renaissance of China and the decline of Europe. Lastly, "Our Revolutionary Era," Chapter 5, gives an overall picture of the character and sequence of revolutionary changes and the yeasty period following them.

In anticipating the future, the idea of a picture puzzle does not only work on a big scale but in any field of interest and endeavor. You decide how big or small the frame of the picture is.

Don't get so far ahead of the parade that people don't know you're in it

THE STARS OF RENAISSANCE

The history of civilization is that things get better. Life expectancy, living conditions, and freedom of choice have improved over the millennia, despite all setbacks and shortcomings. What distinguished the men and women who led the way?

Many of them were like Einstein, willing to detach from the values, rules, and expectations of their time to aspire to higher goals. The price to pay often was envy and resentment, almost a natural consequence when established concepts or knowledge are challenged by breakthroughs and discoveries. The Japanese have a straightforward saying: "The nail that stands up gets hammered down."

Many scientists, inventors, and philosophers who led parades into enlightenment and progress often got hammered down. Over the centuries, many were forced back into the parade of the establishment. Often far ahead of their time, they were getting in the way of those in charge and stumbled over restrictions. Nothing held them back more strongly than religious dogmas, even in times of great intellectual renewal.

In the fourteenth century, one of the most innovative periods in human history began, and was later called the Renaissance. Men and women stood up to break the shackles of the mindset of the Middle Ages and open the door into the modern age. Full of thirst for knowledge, thirst for action, and with great curiosity, they laid the foundation for our modern civilization.

STUMBLING BLOCKS ON THE PARADE ROUTE

The skies are praising the eternal God's glory,
Their sounds carrying forth his name,
Him the world is praising, him the ocean honors,
Listen, o human, to the divine words,
Who carries the skies and the countless stars?
Who guides the sun out of her canopies?

Those are the words of a hymn Ludwig van Beethoven set to music in the nineteenth century. How much more the heavens were in God's hands in the beginning of the Renaissance, when the Catholic Church was in full power and the clergy in charge of science and education, not yet challenged by Martin Luther. It was, however, widely accepted that the Earth was a globe. Eratosthenes, around 250 BC, had calculated what Aristotle and many sailors had speculated, that the Earth was a ball with a circumference of about 25,000 miles. But the doctrine was that it was a ball in the *center* of the universe and the church strictly held to the theory of Claudius Ptolemy, the geocentric (earth in center) system, which Ptolemy had established in the second century AD.

A man stood up against it. Contradicting the official Ptolemaic theory, Nicolaus Copernicus, born in Poland in 1473, worked out an early description of a heliocentric (sun in center) model of the solar system in full mathematic detail: the Earth rotating around the Sun.

Out of fear that his ideas might get him in trouble with the church, Copernicus delayed publication. Only shortly before his death in 1543 did he allow one of his students, Rheticus, to publish his original manuscript, *De Revolutionibus (On the Revolutions of the Celestial Spheres).* But an unauthorized preface was added, downplaying Copernicus's work by saying that the contents were merely a device to simplify calculations.

Twenty years later, in 1564, Galileo Galilei was born. An Italian scientist and philosopher, he had a wide range of talents, including playing the lute and painting. He became the pivotal figure in the comprehension of astronomy and laid the foundation of a correct understanding of dynamics and gravity. He also investigated cohesion (the intermolecular attraction by which elements of a body are held together). Galileo was not only leading the parade in scientific cognitions. He is also credited with inventing the pendulum-clock concept, studying the motion of the pendulum as early as 1582, even sketching out a design for a pendulum clock shortly before his death, but he never built one. We will return to the pendulum clock in Chapter 5, "Our Evolutionary Era."

It's not surprising that Galileo, a universal genius, grew interested in the secrets of the heavens. In 1609, after the discovery of lenses was reported from Holland, he was among the first to use a telescope to observe the sky. He constructed a refractor telescope with a magnification power of 20—amazing at the time but poorer than a modern amateur telescope—and observed the heavens. By finding out that Venus, just like the moon, goes through a full set of phases, he presented the first empirical evidence that allowed a test between the Ptolemaic and the Copernican systems. It proved that the Ptolemaic system was not only more complicated but also incorrect. This discovery shook the foundation of the theological and philosophical worldviews.

Galileo published the chief elements of his celestial mechanics in *Dialogue on the Two Chief World Systems, Ptolemaic and Copernican* (1632). It was supposed to be an objective debate about the two systems. Unfortunately, Galileo put the pope's favorite argument in the mouth of one of the characters and then proceeded to ridicule it. He did not get away with it. It had taken the Catholic Church 73 years to ban Copernicus's *De Revolutionibus*, but in Galileo's case, his famous trial and recantation took place within five months. For the rest of his life, he had to live under house arrest in his villa outside of Florence. The blocks in the road had stopped the whole parade.

"In question of sciences," Galileo said, "the authority of a thousand is not worth the humble reasoning of a single individual." The church could make Galileo stumble, but it could not stop enlightenment over time.

THE TIME TO MAKE THE LEAP

About 200 years later, an Englishman, Charles Robert Darwin, was about to not only shake up natural sciences and religious dogmas but also challenge God. By then the Catholic Church had to a certain extent lost control of science and education. Gutenberg had made the first steps toward inventing his printing machine in 1436 (completed in 1440), which opened access to the written word to average people.

The Bible and its message then reached a vastly wider population, as did scientific literature. Reformation followed in 1520, and in England, Henry VIII further weakened the Catholic Church in 1534 by separating from it and creating the Anglican Church. The French Revolution (1789–1799) shook up France in a call for freedom, equality, and brotherhood. The Industrial Revolution during the eighteenth and nineteenth centuries altered the working and social world—the modern age was well on the way. But one dogma survived all reformations, the creation of the Earth and humankind by God. That dogma was about to be challenged.

As his father had wished, the young Charles Darwin first studied medicine, but soon he was disgusted by the anatomy and the brutality of surgery and instead developed an interest in natural history. His father worried about Charles, thinking him a ne'er-do-well. So he decided to send him to Christ's College in Cambridge to qualify as a clergyman, which was a reasonable career move in his day. Anglican parsons earned quite a good income; this was probably the reason why most naturalists in England were clergymen. Their apostolic mission merged with their ambition to explore the wonders of God's creation.

Darwin finished his studies in theology in 1831 but showed no rush to take holy orders. Instead he spent years of work on the *Beagle* expedition, which allowed him to study geology firsthand. He built up a reputation as a geologist and earned fame as the author of *The Voyage of the Beagle*.

His observations of biology led him to understand the transmutation of species and to develop his theory of natural selection in 1838. Although he was convinced about his theory of evolution, worries that this thesis would associate him with radical democratic agitators who were seeking to overthrow society made him postpone a publication that could mean his ruin. In an effort to discover holes in his thesis, he embarked again on extensive experiments with plants as well as consultation with animal husbanders. He was eager to continue his research and find as much evidence as needed, to meet the disagreement when presenting his thesis.

Fully aware the time was not right, at first Darwin only shared his theory with close friends. He trusted the process to support the best result. In June 1858 the information that Alfred Russel Wallace had a similar theory forced the publication of Darwin's theory.

After being held back for 20 years, *The Origin of Species* was published on November 22, 1859, and 1,250 copies were quickly sold. By common consent, the scientific community established evolution as the dominant scientific theory of diversification in nature. In a national recognition of Darwin's preeminence he was buried in Westminster Abbey, close to Isaac Newton.

A MATTER IN DISPUTE

These historical developments around Darwin prepared the ground for a new understanding of the role of the church, and put in doubt questions once so simple to answer: "Where do we come from? What are we here for? Where do we go?"

The Christian churches, which for such a long time were swinging the baton of the progress parade, had to lay it down. But we are not at the end of confrontation between science and religion. As I will describe in Chapter 5, "Our Evolutionary Era," a big confrontation is on the horizon, a matter in dispute that challenges God as creator and pushes mankind to be in charge of its own species.

Speculations are drifting far ahead of the parade, promising the end of diseases and even the postponement of death, while the necessary education and debate are lagging far behind. The clash between feasibility and justification of genetic engineering—the question of who has the right to lead this parade—will disrupt the scientific world.

In the meanwhile the struggle to handle our daily challenges does not come to a halt, and we keep looking for leadership to show us the way.

THE TOUR GUIDE MODEL

For a long time I have been a just-in-time speaker, parachuting in and helicoptering out. Then my publisher wife, Doris, came along. She left her proper job and became my in-house editor, leaving neither of us with a proper job. That opened a great window of opportunity. Now, when I give a speech, we either fly in earlier or stay on longer, adding some days of exploring to each destination. When I was writing about this parade Mindset, the tourist guides we see everywhere came to mind. They solve the problem of leading and at the same time staying with their parade with a simple and effective method; they wave the follow-me flag to signal their leading role.

The aligning element between tour guides and other leaders is that they have to stay in sight of their parade. Usually a person achieves a leading role for talents and abilities that make him or her stand out. Visionary thinking is one of them. But the daily challenge in business and politics lies not only in the fundamental skills of leadership but also in the necessity to *stay within the field of vision of those you want to lead*. There is a joke within the Vienna Philarmonic Orchestra. Before the performance begins, someone asks the first violinist:

"What is he conducting tonight?"

"I don't know what he is conducting," is the answer, "but I know what we are playing."

Even the most talented leaders need the parade to put an idea into practice. If we leave the parade too far behind and run ahead with our vision, we will be running empty miles. The reality in politics is a little different. In election circles, the average politician is caught in short-term fixing. Some, above that, such as former German chancellor Helmut Kohl, hold on to a larger mission. You will read about him in Chapter 4, "Europe."

HEADING FOR A MOVING TARGET

Keeping in sight of the parade does not get easier when anticipating the future. In the previous mindset section, I explained that *Megatrends* and *Megatrends 2000* were successful because I connected things people sort of knew about to form a picture of the new whole.

It helped me that during my years in politics I happened on some sage advice from Al Smith, the four-term governor of New York who ran for president in 1928: "Don't get so far ahead of the parade that people don't know you're in it." He was a good ol' boy with a razor-sharp mind. One time when he was campaigning for governor, a heckler from the back of the room yelled out: "Tell 'em all you know, Al. Won't take but a minute!" Smith shot right back: "I'll tell them all we both know. Won't take any longer."

Al Smith, as it happens, was leading the short parade of Catholics running for president, although he lost to Herbert Hoover. John F. Kennedy was the second Catholic to run, but he was the first to *win*. John Kerry was the third Catholic to run for president. I see it as a sign of America's maturity that Kerry's religion was not at all an issue during his campaign, unlike the campaigns of Smith and Kennedy.

Al Smith's parade metaphor stayed with me and became one of the best pieces of advice on leadership that I have ever heard. It helped me not to fall into the trap of drifting out too far into the unknown, where anticipating ends and speculation and guessing begin. We care much more about who lives next door than who lives on the next street; the same proximity factor applies to the future. People want to know what is right out in front of them, related to what they know now.

Speculations and interpretations provide a certain appeal in relation to neighbors and the future. Right after the moon landings at the end of the 1960s, many commentators already envisioned the first real estate offices on the moon and adventure travel offered to Mars. Television was going to be the great new instrument of education; helicopters were going to replace commuting cars, pills were to replace food, our households were to be run by robots, and our cars remotely controlled. But all of these exercises in straight-line extrapolation were like balloons let loose into the sky: They soon ran out of air.

THE SENDER HAS TO STAY WITHIN THE RANGE OF THE RECEIVER.

How far ahead of the parade you should be is not easy to calibrate and differs with circumstances. Politics asks that you both be ahead to demonstrate your understanding and compassion for your constituents and be slightly in advance of the times to demonstrate your vision. Corporate leaders have to be only imperceptibly ahead of their parades: They have to attend to business in a no-nonsense way. In technology, it seems that leaders in the various fields are expected to be somewhat out ahead of everybody. In the end, the marketplace will decide: for politicians, the voters; for business and technology, the consumers. For those who would desire to reveal the future, almost all err on the side of being too far ahead of the parade; rein it in a little.

Resistance to change falls if benefits are real

Liberté, Egalité, Fraternité ou la mort!
Réglage, Égalisation, Bureaucratie ce la mort!

"The European constitution is perfect—although maybe a little less elegant than the Constitution of the United States," said Valéry Giscard d'Estaing on April 20, 2005.

On May 29, the French people answered: *"Non!"* The French, and a little later the Dutch, saw no benefit in a constitution written for and by the political elite, 800 pages of regulations and old treaty agreements that would make the European Union neither more transparent nor easier to understand. The "elegant" American Constitution has 12 pages and holds only principles, understandable for anyone. The political earthquake that followed the French and Dutch rejections sent reverberations across Europe. But the no-referenda, which were disasters for supporters of constitutional change, opened a great opportunity for those who want a constitution that serves the people of the 25 member countries of the European Union, all of whom have to approve it.

Not too much has been moving in the European Union since then. More energy went into defending the necessity of ratification instead of the need to change it. The resistance of people will not fall until changes are made and benefits (if there are any, which is still questionable) are apparent. The European Union faces the challenge of responding to reasonable demands of people, like a meaningful constitution, and of establishing understanding for economic necessities such as cuts in unaffordable welfare programs. This is a difficult responsibility for the European Union, which does not yet know who it is and where it is going.

More than 40 years ago, when the European Union was taking its first steps as the European Economic Community, I would not have believed it if someone told me that one day I would live in Europe. I was on my way to Asia to spend a year in Thailand. It was in 1967, during my time with IBM, when I had my first big experience of the importance of making benefits transparent before change can be expected to be embraced. I was in Thailand to develop a project to help accelerate agricultural development in the northeastern part of the country. We tried to interest the farmers in the idea of planting a third crop of rice and also planting other crops between the rows of established crops. But the farmers resisted. They resisted—we found out later—for a very good reason: They knew that their distribution infrastructure could not handle such an increase in productivity. Once we solved the distribution problem, they embraced the new ways of farming and the additional income that resulted.

Ever since then, I have visited Asia several times each year, and I have always been fascinated by the energy that the Asians, especially the Chinese, put into proceeding against all odds and dealing with any change, as long as they were sure of the benefits. During the Cultural Revolution, many left their country and became a strong economic power as Overseas Chinese. They are now coming back as they see the great opportunities returning to the homeland. While Overseas Chinese add their financial and intellectual potential and entrepreneurial verve to the Chinese talent pool, within China millions of the rural population are seeking ways to get out of poverty. We got to know two of them in Shanghai.

The couple Li Fong and Li Chuang grew up in a little village in central China. Li Chuang's father was a vegetable farmer, and from early in life young Li worked with him on the farm as his grandfather and his great-grandfather had done. When he married Li Fong, it seemed clear what was ahead of them. But the talk of good work in big cities had found its way even into this remote small village. Chuang began to speculate that with a better income he could send his child to better schools, support his parents, and at the end, improve life for all the family. The benefits seemed strong enough to fight for permission to move to the city.

Doris and I got to know about Li Chuang several years ago during one of our stays in Shanghai.

We always stay at the Portman Ritz-Carlton, not only because of the wonderful hospitality we experience there but also because the hotel is next door to a typical Chinese neighborhood. Old and new China meet within a few meters. Adjacent to our five-star luxury accommodations, where some suites costs more for one day than most Shanghainese earn in a year, are old town houses with laundry hanging out of windows or simply hung on wires attached to trees or streetlamps. (Every time we go to Shanghai, more of those houses have been replaced by modern buildings with modern equipment.)

Around the corner from the hotel was a hut, a little stand where a couple was selling fruits, snacks, and drinks. Behind the stand was something that looked like an attached tent. We thought it might be a storage room. Doris was especially interested in how the couple experienced the discrepancy between their stand and Ferragamo and Louis Vuitton just 50 meters away. She asked Michelle Wan, the marketing manager of the hotel, who over the years has became a friend of ours, about the couple, and Michelle told us Li Chuang's story.

The Lis had risked a big change in their lives. Their bet had been that in Shanghai, where a vast amount of construction is going on, selling snacks and drinks should be a good business. They built that little stand near the hotel, and because construction in the neighborhood goes on around the clock and their customers are mainly construction workers, their service is 24 hours a day. What we thought was a storage tent was their home, where they take turns sleeping.

Such conditions of living are upsetting to us, but we cannot make the mistake of putting ourselves in their condition. Michelle assured us that they are quite content; the steady stream of workmen coming and going generates enough revenue to sustain them and their child living with their parents back in their village, and a little leftover. Sometime near the end of 2005, Michelle added, she had asked Li Chuang why she didn't see his wife at the stall very often anymore. He told her that she was running their second stall down the road. When we were in Shanghai in July 2006, we bought some peaches at their booth and we noticed that they did not sleep behind the stall anymore.

In China the eagerness for a share in the growing economic pie is driving people to where the action is. What a difference from what I monitor in Europe. The European mindset is upside down—benefits first, and then we'll see. Instead of moving to where the jobs are, many in Europe still expect jobs to come to them, still believe that a job and a house are for a lifetime. I was upset when I recently watched a TV discussion in which young Viennese said they wouldn't move to a job a hundred miles away. They would rather stay unemployed, supported by the government.

Fortunately, those young Viennese don't represent all Austrians or all Europeans. Europe does stand for tradition and constancy, but that does not mean that everything will stay the way it has been and that work and money will be distributed as they have always been, bound to old ways of thinking. The constancy is that we have to make our living, as employers or employees, and the ones who embrace necessary change early on will benefit the most. Again, sports can be taken as a model.

Looking back to the basketball example that I used in discussing the first Mindset, "Most things remain constant," the game stayed the same, but few would argue for a stubborn resistance against Hank Luisetti's new technique of one-hand shooting when the benefits were so clear: winning with it versus losing with the old way of two-handed shots. Nat Holman's desperate statement, "If my boys ever shot one-handed, I'd quit coaching," was proved foolish.

It was the same after the introduction of the Fosbury flop, when Fosbury's coach first tried to get him to switch back to the straddle method. Dick Fosbury's method was a challenge to the conventional thinking, but it broke world records and changed high jumping forever—or for as long as forever is in sports. In sports, results define quickly the path to the future.

In the business world, change might sometimes take a little longer, but in the end the market decides—just as it does in sports.

There are, of course, cases in which resistance to change is the result of stubbornness or ignorance, but people who like to move on in life usually do not resist change just because they can't stand change. On the contrary, people usually embrace change when they perceive that it is to their benefit.

You don't bend down unless something
is worth picking up.

Ask yourself: What is rewarded? What is punished? The score of the game tells you where change results in benefits and where the cry for change is chasing after soap bubbles. Resistance to change can be reasonable or stubborn.

It is the responsibility of those who lead to communicate the benefits of change. It is the leader's responsibility to make it understood by those in whose name changes are made. It is not their responsibility "to get it." They are not the ones asking for change, and they will not support it unless they truly believe they will benefit.

Do not underestimate people. When they resist change—change you think they ought to readily embrace—you have either failed to make benefits transparent or there are good reasons to resist. In that case, instead of lamenting the resistance, look for their reasons for resisting. Notice that in my sports stories, it was the coaches who resisted, not the athletes.

Things that we expect to happen always happen more slowly

A LONG-DISTANCE RUN

It started as a toy. In the fourth century BC, Chinese children were playing with what could be called an ancestor of the helicopter, a spinner made out of a round stick into which bird feathers were put in crosswise at the top. The rotation of the stick between the palms produced enough ascending force to make the spinner lift into the air.

During the Renaissance, Leonardo da Vinci designed several airplanes, including the first helicopter. None of the models would have been able to fly, but the creative value and the engineering thinking was far ahead of time and imagination.

About 450 years later, the German poet Bertholt Brecht wrote a memorial for a compatriot—an inventor and flight engineer.

The Tailor of Ulm

"Bishop, I can fly,"
Said the tailor to the bishop.
"Watch how I do it!"
And up he climbed, with two strange things
That looked like some crude pair of wings,
To the great, great church's heights.

> *The bishop walked on.*
> *"These are nothing but lies,*
> *A man is no bird,*
> *Never will man fly,"*
> *Said the bishop of the tailor.*

"The tailor is dead,"
Said the people to the bishop.
"It was over in a flurry.
His wings are shattered
And he lies smashed
On the hard church's yard."

> *"The bells will toll,*
> *These were nothing but lies,*
> *A man is no bird,*
> *Never will man fly,"*
> *Said the bishop to the people.*
>   BERTHOLT BRECHT

It was Albrecht Ludwig Berblinger, a German, to whom Bertholt Brecht dedicated the ballad of the "Tailor of Ulm." He was a tragic figure

64

in the line of inventors. An orphan when he was 13, he was forced to become a tailor. But his real interest was in mechanics. In addition to his tailoring, Berblinger worked on inventions, one of them being the first artificial leg with a joint.

His most spectacular invention was a hang glider. Against all the sarcasm he had to face, he had finished constructing the first glider capable of flying in 1811 (proven with a model in 1986). But then the day of the first demonstration, supposed to be his greatest, became his worst. Even the king and his sons had come to see him fly. But a technical problem caused postponing. The next day, the king had left but the princes and his brother were still present. The second try failed. A downwind draft and the wrong moment to start the demonstration made the glider crash into the nearby Danube, accompanied by the hoots and laughter of the crowd watching. It was Berblinger's professional and social end.

Otto Lilienthal, a fellow German pioneer, who completed more than 1,000 glider flights on the "heavier than air principal," summarized his own view: "To invent an airplane is nothing. To build one is something. But to fly is everything." He died in 1896, two days after sustaining injuries in a crash of his hang glider.

The outstanding invention of the Wright brothers was an airplane with a motor-powered engine that flew on December 17, 1903. After a very long time of trial and error, the modern history of aircraft had begun, and we are still improving and perfecting it.

More developments are sure to come. For example, it is well known to airframe designers and aerodynamic engineers that the big-span, fixed-wing configurations of today's commercial aircraft are suboptimum, as they need to produce a lot of lift during takeoff and landing because of the heavy load they have to support. During high-speed flight, however, they produce a huge amount of drag that lowers fuel efficiency. Because of aerodynamic principles, flaps can only partly improve this situation. Various concepts for adaptive wings have been around for years, yet the wingspans of commercial aircraft keep getting bigger.

Things we expect to happen always happen more slowly. Even for Arthur C. Clarke, a man whose name is closely connected with the future. Clarke is probably best known for the short story on which the film *2001: A Space Odyssey* was based and for his early anticipation of the use of satellites for communication. In his time line for the twenty-first century announced early in the year 2000, he saw the first human clone created in 2004; the last coal mine closed in 2006; and electronic monitoring virtually eliminating professional criminals from society in 2010.

Not only hot air

The first apparatus powered by steam, just like the first helicopter, was created as a children's toy, following the line of least resistance. The aeolipile was built in 100 BC by Hero, a Greek from Alexandria.

About 1,800 years later, Thomas Newcomen, an English ironworker, invented usable steam engines. The first was installed in a coal mine in Staffordshire in 1712 to pump water out of a mine. Newcomen's engines were extremely expensive but also unbelievably durable; 127 years after it was built, one was still used in Pentlich; another one was working until 1934 in Barnsley. Scotsman James Watt, who is often wrongly seen as the inventor of the steam engine, was improving the technology in the 1760s, making the engine faster, safer, and more fuel efficient.

As the story goes, it was during a walk on Glasgow Green when Watt had the initial idea. Glasgow marks the spot with a stone because, as the Scots say, this is the spot where the Industrial Revolution really began.

The time span from playing with an idea to having usable product has gotten shorter over the centuries but still takes time, and mostly longer than we expect. The computer was no exception.

The first computers were designed in the 1930s and 1940s. Konrad Zuse of Germany is generally credited as their inventor. Zuse, whose company was taken over by Siemens in the 1960s, created the first storage-programmable calculator in 1938. It did not work very well. During World War II, the Third Reich's Aerodynamic Research Institute funded his work, and other models followed. The first computer to match the current idea of a computer, the EDSAC, was built and designed in 1949 at the University of Cambridge, England. In 1974 a new computer species, called a microcomputer, often sold as a do-it-yourself construction set, started to conquer the market and find its way into companies and households.

WORDS LEARN TO FLY

The stories of the new technologies of the telephone, motion pictures, and radio do not differ much. They began their long-distance run in the nineteenth century, lasting through the next centuries to come into our businesses and homes.

In the 1870s Alexander Bell and Elisha Gray independently designed the first devices that could transmit speech electronically. By the 1880s there were almost 48,000 telephones in the United States. The numbers exploded: By 1910 about 5.5 million telephones used the Bell System. In 1924 more than 15 million telephones were counted.

But though the telephone was invented at the end of the nineteenth century, it didn't become ubiquitous in the United States and Europe until late in the twentieth century. Follow-up products increased the speed of diffusion. But AT&T's "picture phone," introduced in 1939, was way too far ahead of the parade and is just now catching on as picture cell phones.

For the fax machine to go from introduction to omnipresence took only 15 to 20 years. I well remember the first fax machine I purchased in 1983. It was a Pitney Bowes and was about the size of a pickup truck. My problem at the time was that there was hardly anyone I knew to whom I could send a fax. If there are only 20 fax machines or telephones in the world, they are not of much good. Another rapid technological diffusion was mobile phones, which took less than 10 years to reach critical mass at which time they really took off.

Pictures Learn to Run

About the same time the telephone started to transmit words over long distances, photography took the huge stride into motion pictures. In France, Louis and Auguste Lumière, inspired by their father's enthusiasm about a demonstration of Thomas Edison's kinetoscope, started to work on a device they called a *cinématographe*. In December 1895 they showed their first motion pictures in the basement lounge of the Grand Café on the Boulevard des Capuchines in Paris. Viewing the premiere of a Louis Lumière film inspired a Parisian journalist to predict a bright future:

Photography has ceased to record immobility. It perpetuates the image of movement. When these gadgets are in the hands of the public, when anyone can photograph the ones who are dear to them, not just in their immobile form, but with movement, action, familiar gestures and the words out of their mouth, then death will no longer be absolute final.

One of the movies of the Lumière brothers was *L'arrivée d'un train à la Ciotat* (*The Arrival of a Train at the Station*). The main effect was fright. The audience is said to have reacted with terror, fleeing from the locomotive as it appeared to bear down on them.

Soon the brothers saw their invention with the eyes of entrepreneurs and opened a theater for exhibiting their films. Lines stretched down the block. Audiences were enthralled. The Lumières trained additional cameramen and sent them on missions around the world. Other pictures soon followed, and motion pictures became a separate industry. Dedicated theaters and companies formed specifically to produce and distribute films, while motion picture actors became major celebrities and commanded huge fees for their performances. By 1917, Charlie Chaplin had a contract that called for an annual salary of $1 million.

The excitement spread. D. W. Griffith, who was one of the important, though controversial, individuals in the development of film as an art, said in 1915 in an interview with the *New York Times*:

The time will come, and in less than 10 years, when the children in public schools will be taught practically everything by moving pictures. Certainly they will never be obliged to read history again. Imagine a public library of the near future, for instance. There will be long rows of boxes of pillars, properly classified and indexed, of course. At each box a push button and before each box a seat. Suppose you wish to "read up" on a certain episode in Napoleon's life, instead of consulting all the authorities, wading laboriously through a host of books, and ending bewildered, without a clear idea of exactly what did happen and confused at every point by conflicting opinions about what did happen, you will merely seat yourself at a properly adjusted window, in a scientifically prepared room, press the button, and actually see what happened.

Griffith's prediction has a certain present-day ring to it, doesn't it?

SOUND AND SIGHT

In the cluster of innovations at the end of the nineteenth century, radio burst on the scene. And as with most innovations, radio began with a series of incremental scientific discoveries and technical refinements that eventually led to the development of commercial applications. I remember the big excitement when radio came to my hometown—Glenwood, Utah. The grown-ups gathered in the parlor to listen to President Franklin Delano Roosevelt's "fireside chats." By the time radio got to Glenwood, it was pretty well established; it had taken its time to get to us.

One of the major pioneers in radio was Heinrich Hertz, who applied James Clerk Maxwell's theories to the production and reception of radio waves in 1888, long before its run to commercial use began. Guglielmo Marconi, a boy with strong interest in physical and electric science studies, picked up the works of Maxwell, Hertz, Augusto Righi, Sir Oliver Joseph Lodge, and others and in 1897 became the first to be granted a patent for a wireless telegraph. Marconi's demonstration of a practical system for generating and receiving long-range radio signals raised interest worldwide. But until 1912 there was no licensing or regulation of radio transmitters in the United States. Civilian radio activities were completely shut down during World War I, between 1914 and 1919, but the wartime consolidation under government control led to important advances. After the war, with the adoption and improvements of the vacuum-tube equipment the radio industry finally expanded when returning to civilian control.

The radio industry finally became profitable and major corporations were founded, including the American Telephone & Telegraph Company, General Electric, and Westinghouse. Because of pressure from the U.S. government, Marconi's assets were sold to General Electric, which used them to form the Radio Corporation of America (RCA). It had taken a long time from the debut of the first radio stations in the early 1920s to having radios in our homes. (Again, Chapter 5, "Our Evolutionary Era: No Next Big Thing," goes into detail about the larger commercial benefits of these revolutionary breakthroughs that were generated through long periods of improving and perfecting the primal ideas.)

The expectations of the new medium quickly rose high as the technology further improved. In 1921, Velimir Khlebnikov, a Russian poet, saw a great promise: "The Radio of the Future—the central tree of our consciousness—will inaugurate the new way to cope with our endless undertakings and will unite all mankind." Eighteen years later, World War II proved that hope to be overly optimistic.

In 1923 radio featured J. M. McKibben's *New Way to Make Americans*. Of the new technology, McKibben said:

> Today this nation of ours is slowly but surely being conquered, not by a single enemy in open warfare, but by a dozen insidious (though often unconscious) enemies in peace. Millions of foreigners were received into the country, with little or no thought given to their assimilation. But now the crisis is upon us; and we must face it without a great leader. Perhaps no man could mold the 120 million people in a harmonious whole, bound together by a strong national consciousness: but in the place of a superhuman individual, the genius of the last decade has provided a force—and that force is radio.

The next year, there was another prediction. *New York Times* science editor Waldemar Kaempffert wrote:

> It so happens that the United States and Great Britain have taken the lead in broadcasting. If that lead is maintained it follows that English must become the dominant tongue. Compared with our efforts at mass entertainment and mass education, European competition is pathetic. All ears may eventually be cocked to hear what the United States and Great Britain have to say. Europe will find it desirable, even necessary, to learn English.

Today, living and traveling around in Europe and hearing almost every radio station playing American songs, I must say that Kaempffert was not all wrong.

RCA introduced the next step in communication, its version of television as the successor to radio, in 1939, at the World's Fair in New York

City. RCA Chairman David Sarnoff announced that "now we add sight to sound. It is with a feeling of humbleness that I come to this moment of announcing the birth, in this country, of a new art so important in its implications that it is bound to affect all society. It is an art which shines like a torch in a troubled world."

However, TV was still years away from having the reach of radio, and the *New York Times* was skeptical. In an editorial after RCA's announcement in 1939, it said: "The problem with television is that people must sit and keep their eyes glued to the screen; the average American family hasn't time for it. Therefore the showmen are convinced that for this reason, if no other, television will never be a serious competition of [radio] broadcasting."

It did not cool down the hype. In 1946 Thomas Hutchinson—a production director for a TV corporation—wrote, "Television means the world in your home and in homes of all people of the world. It is the greatest means of communication ever developed by the mind of man. It should do more to develop friendly neighbors, and to bring understanding and peace on earth, than on any other single material force in the world today." The prose is reminiscent of what has been written about the Internet.

The run-up to having television in our living rooms took a few decades. After TV had been demonstrated many times to many small audiences as early as during the 1920s in the United States, it started in 1932 in England with a cabaret program. In 1933 Iowa State University started broadcasting two weekly television programs with radio station WSUI. In 1936, when only about 200 television sets were in use worldwide, Germany broadcasted the first Olympic Games. In 1948 cable television was introduced in Pennsylvania to bring television to rural areas. By that time, one million homes in the United States had television sets. The first time I saw television was in 1952 when I watched, in a University of Utah fraternity house in Salt Lake City, the games of the World

Series between the New York Yankees and the Brooklyn Dodgers (the Yankees won 4 games to 3).

My favorite elitist television quote for the new mass media is from the year 1963, when the great, elegant poet T. S. Eliot said, "It is a medium of entertainment which permits millions of people to listen to the same joke at the same time, and yet remain lonesome."

In the stream of technological evolutions, uncounted business opportunities floated along. Beside technological enhancements on large scales and often significant consequences, more trivial objects catch the attention of attentive observers of business opportunities. Such a man with an entrepreneurial spirit became the long-term father of an improvement in the production and quality of a beverage without which many would not want to start their day: coffee.

AN OVERNIGHT SUCCESS THAT TOOK 20 YEARS

My wife says I make the best coffee in the world. Under these circumstances, you will understand that I was shocked when her daughter bluntly called the very same coffee undrinkable. Her devastating damnation was caused by her addiction to coffee produced by a little coffee machine. When she and her husband built their new home, one of their (now four) coffee machines got its own alcove in the bedroom (where, if we had one, we would have placed the family saint). That's how I got introduced to Nespresso and got interested in the company's story.

By the 1970s Nestlé had pretty well covered the instant-coffee market for consumers. A spoon or two of instant coffee into hot water for those who had no time but much need for a quick cup in place of the "real" espresso. Not perfect, but fast and clean.

Earlier in the late 1960s, one of Nestlé's competitors had the idea of a capsule per cup. The idea was to offer the best-quality espresso, as made in Italian café bars, directly in homes and workplaces. In fear of missing an opportunity, Nestlé bought the patent, which was still in pretty raw form.

The long journey to develop the product began. During the 1970s Nestlé's research and development department perfected Luigi Bezzera's original concept. The machine and the capsules were completed in the mid-1980s. In 1986 Nespresso was founded and the Nespresso system was launched—a technically advanced small espresso machine with pre-measured capsules of ground coffee for preparing espresso at home or at work—into the anticipated growing gourmet coffee market.

But Nespresso, the new Nestlé daughter company, might have never arrived at its goal had the product not had a father, a board member of Nestlé, who firmly believed in the idea over the years and fought for it until it really took off. It was a long, tough journey; the road was so stony that over the years, no competitor, watching Nespresso failing, bothered to step in. Heavy losses, three times at the brink of closing—a road to nowhere, as many in the company said. But the Nestlé board-father did not lose faith in this seemingly lost daughter. After years, modest signs of life began to appear, but it took another 10 years for the real breakthrough. "What was it?" I asked Olivier Quillet, Nespresso's international marketing manager. "The breakthrough came in the year 2000," he said. "That was thirty years after the purchase of the patent. It was the concurrence of a newly designed Nespresso machine in a choice in colors, the first TV advertising campaign, and the launch of more than fifty European outlets." Nespresso, well positioned to prosper in a world of increasing affluence, began its march into homes and workplaces on a great scale. A new unit of a large, highly structured organization had been able to create a successful innovation.

The score of the game: Nestlé Nespresso SA is one the fastest growing units of the Nestlé group. In February 2006 Nestlé announced that Nespresso's revenues had grown 30 percent, its product turnover now more than half a billion dollars, heading for $1 billion within two years. Nespresso now sells in more than 35 countries at more than 12,000 points of sale. Each year new accessories are added; chocolate, cookies, and many other supplementary products are flourishing beside the core business of the coffee machines and capsules.

Elegant boutique stores invite faithful Nespresso drinkers and newcomers alike to test the many different coffees. Capsules are the umbilical cord that never gets cut. Just as the camera constantly needs film and the razor always needs a new blade, Nespresso machines need a constant supply of capsules. It's a classic business model. In Nespresso's case, buy the coffeemaker, and then buy your monthly supply of ground-coffee capsules, now in more than 40 different flavors and blends, with new offerings constantly added. This would not work unless the coffee was really good, something Nespresso works hard to ensure: It imagines its customers as coffee gourmets who are satisfied with only the best. It turned out to be a longer journey than was first imagined, but the destination is a deep-veined gold mine.

EXPECTATIONS ALWAYS TRAVEL AT HIGHER SPEEDS.

A brief look into history reminds us of the time lines of the past. My lifelong experience has been that things we expect to happen always happen more slowly. With inventions, we continually underestimate the time span required from idea to their full realization. The new fields of biotechnology and nanotechnology will be evolving for all the years left in the twenty-first century.

It is the surprises that overwhelm us, such as AIDS and the terrorist attacks of September 11, 2001. As with the human species and other manifestations of nature, almost all change is evolutionary, not revolutionary. Things just take time—almost always more time than we expect.

You don't get results by solving problems but by exploiting opportunities

The reasonable man adapts himself to the conditions that surround him. The unreasonable man adapts surrounding conditions to himself. All progress depends on the unreasonable man.

GEORGE BERNARD SHAW

Becoming governor of California had been his dream for a long time. And anybody who knew the résumé of the Styrian Oak, as Arnold Schwarzenegger is called in his homeland province of Styria in Austria, knew he would make it one day. Sure, some problems might stand in the way of achieving this goal, and solving them would take some years of planning.

It is now 2003, and let's see, the next election for governor is in November 2006, the primaries taking place that spring. Who would be the likely candidates in the Republican primary? What do we have to do from now until then to position ourselves to win that primary? The Democrats will fight among themselves through the spring 2006 primaries, but we have to focus on winning the Republican primary ourselves. The planning has begun.

But still in 2003, petitioners force an election on the issue of recalling Governor Gray Davis. Suddenly, a door bursts open, and Schwarzenegger is ready to walk through, grasping the opportunity. Schwarzenegger put it best: "This historic election has come about because there is a tremendous disconnect between the people of California and the leaders of California."

On October 7, 2003, Californians were set to go to the polls and answer two questions:

Question 1: Should Governor Gray Davis [who was reelected to a second four-year term only the November before] be recalled? Yes or No?

Question 2: Should it turn out that Governor Davis is recalled, who would you want to replace him?

This last question was followed by a list of 135 candidates, including Schwarzenegger. Whoever got the most votes would win. Governor Davis was recalled, and Schwarzenegger was elected governor. During the campaign, Arnold promised that if elected, he would not make any more movies—surely a pledge that won the day.

The key to Schwarzenegger's success was that he had prepared the ground over the years: He was ready. When the opportunity appeared, he didn't hesitate for a moment.

A HOLE IN THE CURTAIN

At the end of the 1980s, a political opportunity on a huge scale was open-ing in West Germany.

It began with a hole in the Iron Curtain. The idea of iron curtains originates from the theater world. After several serious fires had blazed in European theaters in the nineteenth century, the rule requiring an iron curtain was enforced to avoid fire spilling from the stage to the au-ditorium. The twentieth century gave the phrase *iron curtain* a very dif-ferent meaning.

In 1946 Winston Churchill turned it into a geopolitical banner: "From Stettin [north of Poland] in the Baltic to Trieste in the Adriatic an Iron Curtain has descended across the continent." The Communists did grimly defend its existence, but it was the Wall that split Berlin into east and west that became the brutal symbol. Until the very end, the Iron Cur-tain and the Berlin Wall seemed much more likely to remain than to fall.

The breakdown of East Germany and the unification of the two Ger-manys was prepared and enabled by the changes that had been going on in the Soviet Union since the mid-1980s. To avoid the breakup of the Soviet Union, Mikhail Gorbachev had initiated reforms and given up the claim of Soviet hegemony over the Eastern Bloc, while aiming for stronger cooperation with the West. That led to further democratization mainly in Poland and Hungary. Starting in late 1989, Hungary allowed thousands of East Germans to cross its border to Austria, from where they had open access to West Germany. This cut the first hole into the Iron Curtain.

Contemporaneously with the mass escape through Hungary, the op-position in East Germany gained strength. The ouster by party leaders of Erich Honecker, the East German secretary general who just a short time before had announced that the Wall would stand for 50 or even 100 more years, was a catalyst in the breakdown of the socialist regime.

On the western side of the Wall, Chancellor Helmut Kohl had constantly repeated that unification was the goal of his country. Never-the-less, realistic hopes had been low. Just a few days before the fall of the Wall, West Germans were asked about their hopes for a reunited Germany. More than 80 percent said they hoped that Germany would reunite, but only 3 percent believed it would happen in their lifetime.

On November 9, 1989, I flew from Frankfurt to Warsaw to meet with some leaders of the trade union Solidarity. When I got to the hotel, I turned on CNN out of habit and could not believe my eyes. Hundreds—thousands—of people were climbing the Berlin Wall, a wall that since August 13, 1961, the day it was built, had cost the lives of 190 people. Shortly, crowds were streaming across the border, welcomed with enthusiasm and tears by West Germans, who were rushing to this unbelievable scene. It had happened almost out of the blue.

On the morning of November 10, I took the first plane to Berlin to be a witness to this historic event.

Chancellor Helmut Kohl, who had been fighting for unification for such a long time, amazed by the unexpected fall of the Berlin Wall, saw that the occasion could be seized as his chance to bring down the Iron Curtain and reunite Germany. He knew that it might be his only chance, and he immediately took action. On November 28, 1989, 19 days after the fall of the Wall and without any consultation with his coalition partners or Western allies, he presented a 10-step program to the German Bundestag for uniting Germany. But Germany's unification could only take place with the agreement of the victorious powers of World War II.

It was again Kohl, who by clever negotiations, obtained the agreement of U.S. President George H. W. Bush and Soviet President Gorbachev. There was probably just one second in history in which the Soviet Union was willing to hand over East Germany to the West and NATO. Kohl caught that moment. Important factors were his strong-willed, repeated announcements that unification was going to happen, his quick preparation of practical steps for achieving unification, and his great use of diplomatic contacts in Moscow and Washington. Many others couldn't have seized this historic opportunity the way Kohl did.

Looking back, Kohl noted:

When we began our walk to the unification in autumn of 1989, it was like the transition of a high moor. We stood in the water to our knees, fog hid the view, and we just knew there must be a path somewhere. We didn't know exactly where it was. Step by step we moved on and came safely to the other side. Without the help of God we wouldn't have made it.

ENTREPRENEURIAL MINDS

When you're looking for the shape of the future, look for and bet on the exploiters of opportunities, not the problem solvers.

Jimmy Carter, the first entrepreneurial presidential candidate, saw the chance through the primaries and the use of the new media technologies to—for the first time—win enough delegates to win the nomination before the Democratic Party's national convention took place. Now everyone seeks the presidential nomination this way.

There is a huge difference between problem solvers and opportunity seekers. In her excellent book *The Future and Its Enemies,* Virginia Postrel sees the divisions as those who search for stasis and those who embrace dynamism:

How we feel about the evolving future tells us who we are as individuals and as a civilization: Do we search for *stasis*—a regulated, engineered world? Or do we embrace *dynamism*—a world of constant creation, discovery, and competition? Do we value stability and control, or evolution and learning? Do we think that progress requires a central blueprint, or do we see it as a decentralized, evolutionary process? Do we consider mistakes permanent disasters, or the correctable by-products of experimentation? Do we crave predictability, or relish surprise? These two poles, stasis and dynamism, increasingly define our political, intellectual, and cultural landscape. The central question of our time is what to do about the future. And that question creates a deep divide.

This splendid paragraph was quoted by Richard Karlgaard, the publisher of *Forbes*, in his column of May 8, 2006. He identifies Ronald Reagan and Bill Clinton as opportunity seekers by nature. He sees Al Gore and Hillary Clinton as problem solvers. I immediately thought about Hillary Clinton's dedication to solving America's health care problem from the moment she and her husband moved into the White House; Al Gore is now devoting himself to solving the global warming problem. Karlgaard sees George W. Bush as an opportunity seeker who has surrounded himself with problem solvers.

Windows of opportunity are often blown open and closed again like windows in a storm. You have to be ready to grasp them. Louis Pasteur, who discovered that most infectious diseases are caused by germs and with his work laid the foundation for microbiology, said: "Change favors the prepared mind. I was ready."

MISSING THE TRAIN

The victorious march of digital photography began in 1990, when Kodak put the DCS-100 on the market, the first digital camera available commercially. Digital photography was turning the camera market into something new. The slow death of analog photography had begun. In 2004 the market share of analog cameras had shrunk to 17 percent. Whoever had not jumped on the train in time was out of the game.

Big companies with little flexibility are on the side of the losers. A missed opportunity leaves little chance to recover. The traditional Japanese brand Konica Minolta had to step out of the camera business after more than 100 years. The end came on March 31, 2006. Nikon is not much better off. Production stopped in early 2006. Leica, the German brand, barely made it onto the digital train, stepping into the market with the digital Leica in 1998. Insolvency could only be avoided by fresh investment from the stockholders. Between 2000 and 2004, Leica had to reduce its number of employees, and its banks cut its loans. In 2006 sales finally increased again; the new Digital-Module-R will exceed the company's manufacturing capacities, as it did in 2005. The change is quite late but just in time to avoid a big crisis.

The problem of a declining market for a product can't be fixed by improvements to an already obsolete technology. Instead, the effort to do so often paralyzes the capacity to invest in the new opportunities.

Fred Smith, born in Marks, Mississippi, was ready. He wrote a first paper on the idea of FedEx during his studies in economics at Yale. After leaving the U.S. Marine Corps, where he studied military logistics as a pilot, he turned his idea into reality in 1971. Federal Express, which combined the flagging of the postal service with the new importance of time and reliability, began its operations in 1973.

T. Gary Rogers, chairman and CEO of Dreyers Grand Ice Cream, tells the story of how he seized an opportunity when it surprisingly presented itself. After a business failure, he wanted to try again. "I wasn't discouraged from trying again. I felt this joy in having my own company." He had a family, but no income and no savings. One day, as he tells it, he walked into the local office of Dreyers Grand Ice Cream in Oakland, California. Dreyers was a small company with $6 million in sales and 30 employees. Rogers was telling the owner that he was think-ing about buying a franchise, when the owner got a phone call. "When he hung up, he had tears in his eyes. The bank had just turned him down for a loan to expand his factory." Impulsively, Rogers asked him if he had ever considered selling the company. "Not until just now," he answered. Three days later, Rogers said, "I had an option to buy the business for one million dollars. I got a group of investors to back me and took out a loan."

People who hear his story tend to say, "Gosh, you're so lucky."

"For me, "Rogers says, "it's not luck. It's the ability to recognize an opportunity, and to be ready for it."

A footnote to Rogers's story is that he really does love ice cream, says he eats it every day.

Almost immediately after the first democratic government took office, France's Danone jumped into Poland. The opportunity was 38 million consumers starved of high-quality Western products.

It wasn't easy. Danone's lead representative found a bankrupt country and an agricultural system in shambles, with backward collective farms and no distribution system. Beginning in 1990 Danone took one truck-load of yogurt a week to introduce Polish consumers to the company. Salesmen began going door to door to the thousands of small private stores that had opened to persuade them to stock Danone products. Two years later, the company began local production, at the same time creat-ing a modern farming system to ensure a steady supple of raw milk.

Today, Danone is Poland's leader in fresh milk products, including yogurt, with a third of the market. "It was a competitive advantage to be in early," says Jean-Jacques Doeblin, Danone's director of coordination for central and eastern Europe. "We had time to learn the market."

One of my favorite examples of grabbing an opportunity is George Steinbrenner, the boisterous entrepreneur who owns the New York

Yankees baseball team. He certainly knows how to exploit an opening when it presents itself. His old nemesis, the Boston Red Sox, had been negotiating for months to get Alex Rodriguez, the Texas Rangers's star shortstop who in the previous season had been named the American League's most valuable player. The Yankees were not competing with the Red Sox for Rodriguez. They had their own shortstop, the estimable Derek Jeter. The Yankees were not in the picture at all. Three years before, Rodriguez (known as A-Rod) had been signed by the Rangers to a multiyear contract totaling $252 million, the largest deal in baseball history. But now Rodriguez couldn't stand Texas any longer and wanted to be traded.

By February 2004, the Red Sox had failed to wrap up a deal for Rodriguez. Then out of the blue, George Steinbrenner stepped in, and within a matter of days, he had created a very complicated deal and signed Rodriguez—persuading him to switch from shortstop to playing third base!

Steinbrenner, ever the provocateur, had this parting shot: "You think this A-Rod deal is good? I'm about to sign Ty Cobb."

Rick Warren, well-known for his huge best-seller *The Purpose-Driven Life*, gave what amounts to a lot of good business advice in an earlier book, *The Purpose-Driven Church*. He says don't compete for market share. When he was creating his own church, rather than compete in the church-going market for membership, he went outside the market to those who were not going to any church and offered them something that was not available elsewhere—his version of a purpose in life. Starting from scratch in 1980, Warren launched Saddleback Church in Orange County, California, which is now the fastest-growing church in the United States. The weekend attendance there averages 15,000 worshippers, and there are now dozens of branches throughout the country. Rich Karlgaard, the publisher of *Forbes* magazine, says of Saddleback, "Were it a business, it would be compared with Dell, Google, or Starbucks."

What makes the difference? How to seek and find the opportunities? You remember what Einstein did: He picked the ripe fruits and he made the connections. Once they are made, they often seem quite simple.

In 1970, the Japanese economy was beginning to take off, and the Japanese were becoming wealthy. The 1970 world's fair was held in Osaka, and among many others, John Amos, one of the founders of the U.S. health insurer Aflac (yes, the one with the duck) was visiting. He noticed the surprising affluence of the people and he also noticed how many of the Japanese wore masks to protect themselves from getting the flu or a cold. Many other people probably noticed this, but he saw "the fruits" and made the connection: Japanese are wealthy but risk-averse. What more could an insurance company hope for?

In short, it took four years to get a license, but John Amos's sense of opportunity paid off. Today Japan accounts for more than two-thirds of Aflac's $14 billion annual revenue.

Nephew Dan Amos, who succeeded John Amos as CEO, was apparently blessed with John's opportunity gene, recognizing ripe fruits: The Japanese had been neglecting the high potential of talented women in Japan's male-oriented society. Today more than half of Aflac's 3,300 employees in Japan are women and the company benefits in two ways. "It would be tough to hire the best males because most of them want to work for blue-blood Japanese firms," Amos says. And "the best female talent see better chances of becoming an officer with us than with a Japanese-owned company."

As it happens, I was at the world's fair in Osaka in 1970 as well. It was Japan's first world's fair, and I had been very curious about how it would be done. So I arranged to go there. I noticed two things: (1) the Japanese made sure they had by far the best pavilion, and (2) they made sure that almost all Japanese experienced the winner of this competition.

The government, aware of the opportunity to give the population a boost in self-confidence and optimism, had organized tour groups from all its cities and rural areas. It turned the fair into a kind of solidarity jamboree, and everywhere I saw large groups, all of them wearing special-color group hats, following guides with the group flag, bearing witness to the rebirth of Japan.

One of my own experiences acting on an opportunity arrived early after I had left IBM to start my own company. On the evening of January 17, 1968, I sat in my living room watching President Lyndon Johnson's State of the Union message, in which he declared war on poverty. The president talked about employment for all, including those who then were called the "hard-core unemployed." What was really meant were those in the country's black ghettos who had never been employed. I thought the British definition of the hard core at the time characterized them just right: those who couldn't get jobs even in good times. The president went on to speak of creating a job training program for the unemployed, including the hard core.

As I sat there listening—and having spent time in Washington—I realized that probably neither the president nor anyone around him or anywhere in the government had any program for training the hard-core unemployed. What was needed was a manual for training the hard-core unemployed. I immediately began drafting a proposal to do that, a proposal that emphasized the importance of involving the hard-core un-employed themselves in the process. My company, the Urban Research Corporation, had been founded to help corporations deal with the urban crisis of the day, and this fit right into our mission. I finished the proposal the next day and left the same night for New York to meet with Ford Foundation people. Although I knew how much such a manual was needed, I also knew that if I went to the government for funding, it would take forever to get an answer.

The Ford Foundation was the place to turn to, and the second day after the president's speech, I was in their offices presenting my pro-posal. Among their countless projects, they had not yet focused on how to train the hard-core unemployed, but they had all heard the president's speech.

My proposal was funded within 24 hours, the fastest a Ford Foundation proposal had ever been approved, a measure of the urgency and promise we shared. It became my new company's first big project. A year later it resulted in the production of a 12-volume training manual that was distributed by the Labor Department. President Johnson's term ended the next year, and with his leaving, the energy and dollars going into that particular war on poverty receded.

The entrepreneurial mindset is all about this impulse of seeing and then doing something creative with a presented opportunity.

Many success stories are about men and women in their 30s or younger. But there is no age limit for a new start—adding experience to abilities. Sometimes such a new beginning starts with an end, taking a little shove to make a decisive step. I'm convinced there are many such cases we have never heard of, cheering as they are. But there is one I can tell: the story of Maxine Martens.

I met her in Milan, where I gave a speech at the Milan Fashion Global Summit. During the lunch break a woman came to me with a question, and we started talking. I found her job very interesting—headhunter for executive positions in the fashion world, which she did very successfully. As I learned later, she does her work for global brands like Cartier, Louis Vuitton, Ralph Lauren, Gap, and many more. It turned out she would be in Vienna soon, where my wife and I got to know her better, an elegant lady with a warm heart, and our friendship began. Now we know her story, which she allowed me to use in this book. And I don't think there is a better way of doing so than passing her letter on to you: the story of an end that became a new beginning.

Dear John,

How did it feel when I was fired? It was a landslide which swept away my life. Suddenly the world became a different place.

"You don't fit in to our culture; you aren't driving the business where we want to go. We are restructuring the organization; let's mutually agree to part ways." How could anyone not appreciate my contributions, my personality, my leadership! I fell into a chasm.

Why me? I was still the same person I had always been, a managing director and major contributor in an American search firm. Now I was fired, dismissed, rejected, and no longer needed, wanted, or desired. But there was one statement which captured it all. *"I don't know how to make you feel like part of the team—you don't golf."* While it was said in jest, as an executive recruiter I have learned that this is when people really tell the truth. These few words changed everything. I knew I had to go because our difference and core values were all wrapped up in the perception that being a golfer would make a difference.

After days in shock I decided to breathe. Wasn't there also freedom and opportunity to assess who I was and what and with whom I wanted to work, not just to think about the pragmatic problem of how to earn a living and pay for my daughter's university tuition?

So a week after I was fired at age 55, for *"not fitting into the culture,"* I decided to do what I had been doing best since 1972. I traveled. Since my first buying trip to Europe, I have enjoyed and embraced meeting people whose lives, histories, and life experiences were different from mine.

During my trip I spoke to supporters and detractors, people who wanted to partner with me and those who were only interested in the business I could bring to them. We discussed how our styles and practices would mesh, but not what opportunities we could exploit and build together. The second week I met a Swiss lawyer who, after listening to me, asked: *"Why do you once again want to be part of someone else's vision? Why don't you do it on your own?"*

I was in the business for 17 years. For the last 10 years I had thought about having my own firm, travel the world and meet people, while earning a living pursuing a passion of building bridges between business cultures and companies in America, Europe and Asia, finding and placing individuals who could embrace a company's culture and heritage, adding value by bringing in their own business acumen, experiences, creativity and point of view. But I had not had the guts to make the step to be on my own.

That night in Zurich, I began to picture my own company: an American search firm based in New York. I made my list of how, 55 years old, I wanted to live and work. In the morning, I emailed it to a business friend in Paris. She replied with one message: "Yes, that's you! Do it!"

In the beginning we worked out of my New York apartment: my business partner of 12 years; a young woman of 20; our search administrator, a talented 24-year-old young woman from Vienna; and our operations and financial officer, my son, a recent Georgetown graduate. None of us had ever set up a company and most of us were Americans who had lived and worked in other countries and cultures. Our mission statement, our values, and our philosophies were written down, based on our ideas of the environment and business we wanted to create and the feedback we had received from clients to it.

We found our beautiful office with a wonderful view and within walking distance of most of our clients. Thanks to our lawyer: The day we legally incorporated the company, he asked whether we had found our space yet. When we said, "not 100 percent," he said, "Run, do not walk, over to meet this realtor, he has the perfect place for you and it just became available."

Today we have tripled the original size of the team and the number of clients. We continue to learn, to make mistakes, to seek the advice and to meet wonderful new clients and candidates. We have changed, and we still use many of the skill sets we had before. It is delightful to watch how we are all evolving and changing, just as the clouds which are crossing now behind the magnificent towers on Central Park West.

Maxine

Having started this discussion of Mindset #9 with a quotation from George Bernard Shaw, I would like to end with him:

"People are blaming their circumstances for what they are. I do not believe in circumstances. The people who get on in this world are the people who get up and look for the circumstances they want, and if they don't find them, they create them."

CHANGE IS THE FATHER OF INNOVATION.

Opportunity seekers know the future with its changes delivers the opportunities. Problem solvers necessarily are dealing with yesterday.

Times of change are times of opportunity. When the relationships of people and things are shifting, new juxtapositions creating new needs and desires offer new possibilities. Keep your eye on those who grab openings and do something with them.

Don't add unless you subtract

LEVERAGING QUALITY

I was introduced to the concept of don't add unless you subtract by Francis Keppel, who was President Kennedy's commissioner of education (now a cabinet position, made so by President Johnson). At the time, I was assistant commissioner for public affairs. Frank was an amazingly talented guy. In the 1950s when he was only 29 years old, Harvard's president, James Conant, named him dean of the School of Education—and he didn't even have a Ph.D. At Harvard! The school was no more than average, and Frank's job was to fix it.

Faculty members were prone to add courses to inflate their importance and paychecks. To strengthen the curriculum, Frank laid down a new rule: You can't add a course unless you subtract a course. The rule was designed to make the faculty think twice about the quality and relevance of a new course and about which of their weaker courses they were willing to give up.

Applicable to all fields, Keppel's rule has maintained a role in my life. In the process of writing this book, I began with 27 mindsets that have over the years helped me think about the future. That seemed like too many. So I asked myself which of the 27 had helped me the most. I chose 10 and began to write about them. In the process, I added some mindsets, dropping others that began to seem less important. I finally arrived at my list of 10 but added one last mindset that just had to be included, for a final count of 11. Here are some that didn't make the cut but have served me well:

Look at what is rewarded and what is punished. In both societies and personal relationships, noticing what is rewarded and what is punished can tell you a great deal about people and social arrangements.

A proposition doesn't have to be true; it just has to be interesting. This idea from the great philosopher Alfred North Whitehead is not for general application but is a good way to stimulate thought.

To appraise the viability of a society or a company, examine its ability to be self-correcting.

Growth is regrouping at a higher level. People, institutions, and societies are growing if they continue to reconceptualize their role or mission to more demanding levels.

As with other Mindsets, I am turning to sports where "don't add unless you subtract" has been institutionalized. In all sports, the roster has a cap, and you are not allowed to add a player unless you trade a player. Professional basketball limits each team to 12 active players. If a team wants to add a player, it is forced to think about whom to let go—thus over time strengthening the team (if its judgments are sound). The business world, for good reasons, cannot have that rule, but too often the motto appears to be "Add, add, add." Add more products without dropping weak ones, more personnel without letting anyone go.

3M has a policy of adding and subtracting products each year. Its vision is to be the most innovative enterprise and preferred supplier in every market served. Each year 30 percent of 3M's sales are derived from new products, introduced within the previous four years.

Jack Welch had a rule at GE that each year, each unit had to let go of the bottom 10 percent of employees, which could be replaced by more promising candidates.

This Mindset helped GE to stay efficient and applies in many fields to stay focused. Once we settle on the manageable number of projects, stocks, trends, or boyfriends or girlfriends, the "don't add unless you subtract" mindset will improve quality.

ADD SHRIMP, DROP PORK!

A restaurant in Saigon (now officially Ho Chi Minh City, but never mind) that has become a favorite for my wife Doris and mine is an entrepreneurial celebration of the "don't add unless you subtract" Mindset. When we first went to the restaurant, we did not know what to expect. As we walked in, we saw that the guests were a mixture of locals and tourists. We were seated at a table in the front garden under palm trees and tropical plants. There was an English menu, but it was a little hard to read. When we asked the waiter to help us, he invited us to follow him. "Have look!" All lined up along the walls around the large central sitting area were little kitchens—all in all about 20 rather primitive mom-and-pop kitchens. Each of them was preparing its own Vietnamese specialties, and it was easy to see that some of them were in bigger demand than others. And all contributions were listed on the menu card.

After placing our order by pointing at what we wanted, we went back to our table. A little later, trung chien voi cua (shrimp omelette), muc don thit (squid filled with pork), com chien (roasted rice in a pan), and chuo nuong (banana cake with cashew nuts) were served. The food was delicious, but we backed off from finishing with ruou ran (rice wine with a small cobra floating in the bottle), no matter how much good it would have done for us.

Later, the waiter told us that there is a backlist of kitchens wanting to join the restaurant. The owner of this private restaurant from time to time chooses a new kitchen but only if he subtracts an old one for whose specialty the demand has been declining. This process stimulates a constant increase in the general quality of the food one can order for a meal. Unlike the fast-food centers in U.S. malls where each kiosk is leased for a period of years regardless of performance, the competition for quality is built into this Vietnamese restaurant. We have found that in many parts of Asia, capitalism and entrepreneurialism are not only embraced but also improved on, such as taxis in China.

In America, Europe, and most other places, you pay a certain price despite the quality of the taxi. In China, a bigger and better car goes for a higher price. It starts with small-size 1.2 taxi (indicating the amount you pay per mile), printed on the side, and ends with 2.0, which meets the quality of a medium-size sedan.

CEMETERIES OF INFORMATION

Never go shopping with a hungry stomach is an old rule. We end up with piles of food in our cart that we won't ever be able to consume.

When it comes to feeding our intellect, we sometimes act the same. We walk through supermarkets of information, which overwhelm us and in which it gets harder to select what we really need. Our goal should not be to create cemeteries of information, but cradles of knowledge and inspiration.

In my field of interest, collecting information can easily get out of hand. In the euphoria of my student years, I added constantly—my books were climbing up the walls, filling every space available and more. But that did not necessarily add to my knowledge. So I now have a rule: my library should hold 4,000 books—that was the amount I came to after sorting out the ones that were filling space but out of demand. Since then I have not added a book unless I selected one to give away, stabilizing at least the quantity and, in time, certainly raising the quality and relevance of my library as well.

NEVER THROW MORE BALLS THAN YOU CAN JUGGLE.

The 11 Mindsets of this book should support your decision on which 5, 7, or 10 categories of information you judge to be the most important to you and your field of endeavor. For one person to keep track of all the technological evolutions and geopolitical developments going on today is not possible. Focus on what really meets your needs and interests. Don't add unless you subtract. If you decide to keep track of seven and something comes along that you feel you must keep up with, drop one of the seven that you or the world has lost interest in. The churning of your changes will be a reflection of a dynamic world. To this I would add an aphorism from Lao Tzu: "To attain knowledge, add things every day. To attain wisdom, remove things every day."

Don't forget the ecology of technology

AN INVASIVE SPECIES

A gray blanket covers Australia. It is not a weather front from the Southern Ocean. It came in 1859 from the west, had four strong legs and two long ears, and was brought into the country initially to entertain sporting gentlemen. Thomas Austin released 24 wild rabbits on his property at Barwon Park, southwest of Melbourne. In 1866, only seven years after the rabbits' immigration, 14,253 rabbits were shot for sport alone in Austin's park. A resounding success!

How sweeping, he could never have imagined. The laws of fertility shifted Australia into the position of Goethe's "The Sorcerer's Apprentice," whose broom turns from a blessing to a plague.

See him toward the shore his racing
There, he's at the stream already,
Back like lightning he is chasing,
Pouring water fast and steady.

Once again he hastens!
How the water spills,
How the water basins
Brimming full he fills.

At the peak of the feat, the picture turns. The broom gets out of hand, and splitting it in two parts only makes the situation worse.

Woe betide me!
Both halves scurry
In a hurry,
Rise like towers
There beside me.
Help me, help eternal powers!
Off they run, till wet and wetter
Hall and steps immersed are lying.
What a flood that naught can fetter!

Sometime in the 1950s in Australia, a man shot a rabbit on farmer John Robertson's property. He was fined 10 pounds. A few years later, Robertson's son had to spend 5,000 pounds a year to control rabbits.

Lord and Master, hear my crying!
Ah, he comes excited
Sir my need is sore
Spirits I have cited
My commands ignore.

Australia's call for the master is loud, but the hex to command the uncontrollable forces has not been found, and each intervention causes other disturbances. The gray blanket is still growing.

NEW BROOMS DON'T ALWAYS SWEEP WELL

In Australia, short-term benefits were wiped out by long-term damage. What is true in nature applies to technology. Here and there, the consciousness of the changes to the habitat and the relationships within it is lacking.

And yet this is nothing new. The advances of technology have always resulted in social change. The discovery of fire led to warmth, better food, and the beginning of real community. The wheel, electricity, and the automobile all dramatically changed our social arrangements. The difference today is that the accelerated rate of the technological change has been so great that the *social accommodation* to new technology has lagged further and further behind. The evolution of technology is now running ahead of cultural evolution, and the gap is increasing.

THE QUESTIONS NOT ASKED

Is technology like the good-news, bad-news of rabbits into Australia? The introduction of a new species and the introduction of a new technology never happen in a vacuum. But despite their importance, the consequences of our relationship with technology are not given much consideration. It is our most unexamined relationship. With the advent of a new technology, questions need to be asked:

What will be enhanced?

What will be diminished?

What will be replaced?

THE BOOMERANG OF TECHNOLOGY

Not everything that sails along under the flag of a technological phenomenon really is one. Two of the most significant examples are the Internet and mobile phones. They are not about technology; they are about connecting people. They have, apart from all other aspects, changed the way we communicate. And as often happens, some consequences are as unintended as what happened to the art of letter writing.

In the beginning of the last century, James Joyce handwrote a letter to Nora, who would later be his bride:

15 August, 1904

My dear Nora,

    It has just struck me. I came in at half past eleven. Since then I have been sitting in an easy chair like a fool. I could do nothing. I hear nothing but your voice. I am like a fool hearing you call me "Dear." I offended two men today by leaving them coolly. I wanted to hear your voice, not theirs.

    When I am with you I leave aside my contemptuous, suspicious nature. I wish I felt your head on my shoulder. I think I will go to bed.

    I have been a half-hour writing this thing. Will you write something to me? I hope you will. How am I to sign myself? I won't sign anything at all, because I don't know what to sign myself.

More than a hundred years later, via e-mail, it might read like this:

Nora,
quick thought: got in @ 11.30—since then just hanging round like an idiot *lol*, doin nothing. just hear your voice going <dear> :-(. annoyed 2 clients, skipped appointments. wanted 2 b with u!!!!!!!!!!! ur good 4 me, u make me feel better, less depressive, less

monster-like ;-) , want your head on my shoulder. NVM, will go 2 bed soon been writing this for like 10 mins now. write back soon!!!!!!! hope you will!!!!!!!! never know how to finish these things. CU soon.

THE COMPUTER AND THE POET

Our switch from high touch to high tech, from handwritten and heart-felt to computer-driven, has a lot to do with the glorification of the computer, the cure-all of our time. Sure enough, it will fix our shortfall in educational achievement. America's elementary and secondary schools are in decline? Let's put a computer—that great icon of the technological age—in every classroom. That should fix it. The cry for a computer in every classroom has been with us for a couple of decades. But have we examined what a computer is going to contribute to the real purpose of education? The real purpose of education is learning how to learn, the only way we can continue the lifelong adventure of learning. But we keep pressing for the technological fix without understanding its educational value.

I favor a computer in every classroom. I favor a computer for every child; there is no question about that. I want kids to be friends with computers, to experience computers as collaborators. We could not afford not to push in that direction. But my campaign is to also have a poet in every classroom. What is happening today in America is that most of the energy and money are going to the computer, many times at the expense of poetry, art, music, and the rest of the humanities. But what is needed is technical understanding and artistic imagination. As I write in Chapter 1, "A Visual Culture Is Taking Over the World," we are moving into a world in which the high touch of a product, its design and poetry, will become more and more the way to differentiate it from the products that have more or less the same technological level.

My metaphor of the computer and the poet stands for my belief that we have to nourish not only the intellectual skills but also, in balance, support and feed emotional and spiritual talent. We very much need to keep them in balance—in high-tech, high-touch balance. The more technological our world becomes, the more we need our artists and poets.

WAITING TO BE RESCUED

Once upon a time there was a world without cell phones. No melodious ringing in schools, seminars, movies, gyms, or supermarkets. Then technology endowed us with the possibility to wirelessly connect anytime, anywhere. Nokia, the most successful producer of cell phones, wrote on its flag: "Connecting People." Technology is on the fork of blessing and plague, and again, very few examine it.

Earlier this year I was at Boston's Logan Airport. The plane was late, the departure postponed. My fellow sufferers and I were sitting in a public area waiting to finally board. Across from me was a young family—a father, a mother, and two children. They were not talking too much, at least not the young father, who looked completely bored. Suddenly his mobile phone rang. He straightened and lit up like a Christmas tree. "Hello, hello," he answered, much too loudly.

As I looked at him, I realized that he was being *rescued*, rescued from his boredom, and rescued, not so incidentally, from his family responsibilities. This young father, it seemed to me, has a dysfunctional relationship with technology. As I sat there as a witness, I thought that there must be people all over America waiting to be rescued from whatever was in front of them, rescued by—anybody.

FROM KING TO SUPPLICANT

In the parade of progress, stationary telephone systems were not left behind. During the 1990s, the cornucopia of technical advancement was spilled over all of us. A gentle voice, conducted by a technology-driven, cost-cutting management, usually starts with the heartwarming affirmation "Your call is important to us. Please hold while we ignore it."

That is, of course, only the overture.

"To listen to our 112 menu items, press 1. For the latest running total of the number of customers who say they would rather die than do business with us again, press 2. For someone who is a very nice person but doesn't have a clue, and in any case is on maternity leave, press 3."

After 11 minutes and 33 seconds, the finale comes in a crescendo: "The person you are trying to reach is either on the other line or not in the office. Please try again later."

That rips it.

Those companies call the customer king and treat him worse than a supplicant. I say get rid of voice answering systems immediately. They are offending customers and putting them in telephone hell. Replace those machines with friendly high-touch operators. Then corporations will see where the real cost efficiencies can be made.

I urge any CEO whose company has a voice answering system to call his company and see whether he can get through to himself. That eye-opening exercise would hopefully trigger great entrepreneurial opportunities to create a useful customer service system that doesn't turn into an infuriating maze with no exit.

Again, we have to think more about the ecology of technology.

A RISKY CHOICE

Some applications of technology, on the other hand, are quite amusing. A while ago I read about a restaurant in Paris that blends high tech and high touch into a great mélange. Its name is La Connivance, and it is next door to the stock exchange. In harmony with its environment, the laws of supply and demand determine the price of the meal. When you place an order, a computer instantly adjusts the price to reflect demand. Popular dishes—like popular stocks—go up in price, while less popular ones decline. Now, you can pay the price at the time you order—or you can play the futures market, agreeing to pay the price quoted when you pay the bill.

I thought that was innovative, but when my wife and I talked about La Connivance with a friend, he told us about a bar in Oslo that goes a little further. The guests bet on the same speculation principle, but randomly, a stock market crash is added, dropping the price of drinks to near zero, resulting in a great crunch at the bar. Well, these are only manifestation of what computers can do for us.

SIZE MATTERS

In the process of introducing new technologies, we often forget to think about human scale. Back in the old days in Utah, no one was worried about human scale. It had been taken care of by the leadership of Brigham Young. When he arrived with his wagon train of pioneers in Great Salt Lake Valley in 1847, one of the first things he decided was the optimum size of a parish (or "ward," as the Mormons call them): 300 men, women, and children. Because of the large number of children Mormons used to have, that amounted to only about 50 families. Everyone knew everyone else, all could take care of everyone. The social net was braided with comprehensibility. Human scale.

In most companies, human scale was set aside during the industrial era. We are now returning to it. Gore-Tex is a company with fabric and medical sales of almost $2 billion that has not lost touch with human scale. It limits the occupancy in each of its buildings to no more than 200 people—and each building is no more than two stories tall. With fewer than 200 people in a plant—or office—everyone can know everyone else. Everyone can know what each is contributing to the whole, in contrast to the 5,000-worker plants of the industrial period. The right human scale gives us efficiencies and allows creativity to flourish.

THE HIGHER THE BETTER?

The skyscraper competition of world cities is one of the measures of human scale—or the lack of it. Chicago beats New York, Singapore beats Chicago, Kuala Lumpur beats Singapore, and the list goes on. But it was and is questionable whether skyscrapers are within the considerations of human scale. Is stacking up people for 100 stories human scale? I don't think so. Maybe over time the tragedy of the terrorist attacks of September 11 will lead us to think about scale and its relationship to human beings and desirable human environments as well as about symbolism.

THE BLIND RUSH

Many companies, driven by a fear that new technologies will give competitors an edge, are driven from one upgrade to another. That costs money—more than they think. We often get obsessed about technology at the expense of putting money and energy into the human side of the business. IBM's 360s, the first mass-produced mainframe computers, sold like hotcakes beginning in the 1960s because a mainframe was considered a must for every company. The very unofficial mantra of the

IBM salesmen was that their customers were buying out of "Fear, Uncertainty, and Doubt," or the FUD factor as they used to say. "Right or wrong, we have to do something" tended to be the position. A spreading panic declared that to be competitive in the new technological age, a company had to get itself organized with a mainframe. There was a lot of mindless spending in the hope of not being left behind, just like today.

TECHNOLOGY IS THE GREAT ENABLER,
BUT NOT IN A VACUUM.

Technology is a great enabler, but only when in balance with needs and skills and our human nature.

Each stone thrown into the water has rippling effects; each new technology has consequences that are seldom explored. When a new technology is introduced, make it a rule to ask: What will be enhanced? What will be diminished? What will be replaced? What new opportunities does it present?

PART II

PICTURES OF THE FUTURE

Part II will introduce you to five pictures of the future I have framed. They are the result of putting single pieces together until the puzzle picture is completed.

The pictures I describe allow you to recapitulate how the single pieces, which at first can seem unrelated in connection with others, are forming a whole. Some of the Mindsets which have supported the process are pointed out; other Mindsets that appeal especially to you might come to your mind along the way. At the end of each chapter, examples of Mindsets at work should get you started on your own process of forming pictures of the future in your field of interest.

Chapter 1

A Visual Culture Is Taking Over the World

A farewell to Gutenberg's world of scripture, a welcome to MTV's world of vision? There is unprecedented visual assertiveness in the world today, from art and architecture to high-end fashion and the design of common goods. It is an MTV world, a world where visual narrative is overwhelming literary narrative.

The history of the written word goes back 6,000 years. Words tell, words explain, words advise, words change, words win, words form, words entice, words touch.

> *How beautiful you are, and how pleasant you are, O beloved girl, among exquisite delights! Your stature is like a palm tree, your breasts are like date clusters. I have said, "I will climb up on the palm tree that I may hold of its stalks of dates." And please, may your breasts become like clusters of the vine, and the fragrance of your breath like apples, and your mouth like the best wine that is going down smoothly for my beloved, flowing gently through the lips of those who fall asleep.*
>
> *I am my beloved's and towards me is his desire. Do come, O my dear one, let us go out into the field, do let us lodge along the henna plants. Do let us rise early and go to the vineyards; let us see whether the vine has budded and its blossoms have burst open, and whether the pomegranate trees have bloomed. There I will give my love to you. The mandrakes have given forth their fragrance, and over our doors there are all sorts of the choicest fruits. The new ones as well as the old, which I have saved up for you, my beloved.*

These are words from the Bible, expressions of love preserved in "King Solomon's Song."

Poetry has been in the hands of sovereigns and politicians Solomon, Caesar, Nero, Mao, Che Guevara, Havel, Pablo Neruda. Words express feelings, open doors into the unknown, create pictures of worlds never seen, allow adventures never dared; Jules Verne, George Orwell, Hemingway, Melville. Searches our soul, Proust, Musil, Joseph Conrad. Storytellers, poets, and singers were respected in all cultures.

"The greatest thing by far is to be a master of the metaphor," Aristotle wrote in his *Poetics* in 332 BC.

RUNNING OUT OF DEMAND

But now, the romance is fading. Imagination is outperformed by the instant picture. In a triumphal march, movies, TV, videos, and DVDs are replacing storytellers and books. It is a visual culture embedded from childhood, and this culture is taking over the world—at the expense of the written word. With it, the novel, the cradle of fantasy, is not dead—as has been announced so many times—but it is losing blood at an alarming rate.

"I have no faith in the survival of the novel; it is almost over," said Sir Vidia Naipaul, the 2001 Nobel laureate for literature, noting that the novel has been "dying" for at least a century.

Tunku Varadarajan, the editorial features editor of the *Wall Street Journal Europe*, puts Naipaul's remarks in a more optimistic perspective, at least from the novel's point of view: He wonders if Naipaul is a "tad blinkered" and that maybe what he really means is that nobody writes novels as good as his own. Varadarajan's response to Naipaul invokes Graziano early on in the *Merchant of Venice*:

"I am Sir Oracle,
And when I open my lips let no dog bark!"

Whether it is a lack of attraction or a lack of quality, the result is the same: Our literacy, and with it our verbal and communication skills, are in decline.

The only category of novel that is gaining ground in our increasingly visual world is the graphic novel. Barnes and Noble, America's biggest bookseller, now has separate sections in its bookstores to promote new graphic novel titles. In spring of 2006 I talked with James Killen, the graphic novel buyer for B & N. He reviews and decides which and how many new graphic titles B & N will buy for all its stores. For years we have seen comic book adaptations of famous novels (notoriously used by college students in lieu of the real thing). But now the big sales are in original graphic novels and narratives. The breakthrough, Killen told me, was in 2003 when *Blankets*, by Craig Thompson, was published, a memoir of childhood running to 600 pages. I asked Killen how he would characterize the growth of graphic novels and narratives. He said, "explosively exponential."

"Where is it all going?" I wanted to know. "How high is the sky?" The 9/11 Report will be released as a graphic narrative in the fall of 2006.

The question of how serious a problem for conventional novels was partly answered by the fiction judges of the 2004 National Book Awards when they nominated Christine Schutt's novel *Florida* as one of the five finalists for the annual prize. It sold fewer than 200 copies. That's not all: Three other finalists for candidate for the best fiction by an American author in 2004 had sold between 700 and 900 copies each. The fifth candidate, Kate Walbert's *Our Kind: A Novel in Stories*, sold about 2,500 copies.

The competition is sponsored by the National Book Foundation, a nonprofit organization mostly supported by publishers and booksellers; the National Book Award is their flagship activity. But as Larry Kirschbaum, the longtime chairman of Time Warner Book Group put it, "We are completely closing ourselves off from the culture at large. We are supporting our demise."

After years of explosive growth in the number of U.S. book titles, in June 2006 publishers reported a decline from the previous year. In 2005, the number of book titles dropped to 172,000 from 190,078 in 2004.

THE CONSEQUENCES

In July 2004, the National Endowment for the Arts released a study "that documents a national crisis," as its chairperson, Dana Gioia, said. From 1982 to 2002, literary reading declined among whites, African Americans, and Hispanics, most strongly among Hispanic Americans, dropping by 10 percentage points. By age, the three youngest groups saw the steepest drops. The rate of decline for the youngest adults, those age 18 to 24, was 55 percent greater than that for the total adult population.

"America can no longer take active and engaged literacy for granted. As more Americans lose this capability, our nation becomes less informed, active, and independent minded. These are not qualities that a free, innovative, or productive society can afford to loose. No single factor caused this problem. No single solution can solve it. But it cannot be ignored and must be addressed," summarized Gioia.

The history of civilization is a history of communication. If communication shifts from word to visual, we need to learn a new language to interact. The competitive edge in a visual world will be education in technology and the arts—a technology-driven brain and a lyrical heart. We need the computers and the poets.

How can we be sure that we are in a shift from the written word and our own imagination to instant picture consumption—a visual world? In the first part of this book, in the discussion of the Mindset "the future is embedded in the present," I emphasized that change is a result of a confluence of forces. It is almost never just one force, almost never just straight-line extrapolation. If you think you see a shift to something new, I urge you to ask yourself: Is there other evidence that this is happening? Are these other forces pushing in the same direction? In the present case, is there enough evidence to see a visual world—at the expense of the written word—in front of us? In what follows, I deal with eight forces pushing in the direction of a visually dominant world:

1. The slow death of the newspaper culture
2. Advertising—back to "a picture is worth thousands of words"
3. Upscale design for common goods
4. Architecture as visual art
5. Fashion, architecture, and art
6. Music, video, and film
7. The changing role of photography
8. The democratization of the American art museum

1. THE SLOW DEATH OF THE NEWSPAPER CULTURE

I started to read the *New York Times* when I was 19 years old. Ever since then I have not missed a single issue, no matter where I lived or was in the world, albeit sometimes going to great lengths to get it, often entailing long delays. On this matter, I feel like Marshall McLuhan, who said: "Reading your favorite newspaper is like taking a warm bath." Reading my *New York Times* is like taking a warm bath for me. But it seems fewer and fewer care for such an experience.

On November 7, 2005, the *New York Times* announced that its circulation was down 2.6 percent. Its chief executive, Janet L. Robinson, said that "the media marketplace has been challenging in 2005 and we expect it will continue to be next year." Like other newspapers, the *Times* is struggling to hold on to readers and advertisers in the face of increased competition from the Internet.

U.S. newspaper circulation fell another 2.6 percent in the six-month period ending in March 2006, the Newspaper Association of America announced on May 9, 2006. Sunday newspaper circulation fell 3.1 percent in the same six-month period.

Newspaper circulation is falling in all developed countries. In the European Union countries, for example, newspaper circulation has dropped 5.26 percent over five years. During the same period in Japan, newspaper sales fell almost 3 percent.

Newspaper circulation has been falling for years. In the 1960s, four out of five Americans read a paper every day; now only half of Americans do so. The *Washington Post*'s circulation is down 3 percent from 2005; the *Los Angeles Times*'s circulation is down more than 6 percent. Desperation has produced a new kind of scandal. Newspapers are fabricating (to put it gently) circulation figures. Recently caught in the act: the *Chicago Sun-Times*, the *Dallas Morning News*, and *Newsday*, which was found to have nearly 100,000 ghost readers. Newspapers that once published in the traditional broadsheet size are switching to the tabloid layout, and often printing the steamier stories tabloids are known for. Meanwhile, free newspapers are getting a foothold, and the Internet is siphoning off readers across the board.

At this point I have to make a confession. When I moved to Vienna, this great, sophisticated city, I found, to my big surprise, that the *New York Times* was not available anywhere. So for five years, I had a mail subscription to the daily *New York Times*, paying almost $5,000 a year. But now, with such easy excess to the daily Internet version, I decided to have the Real One sent to me only on Sunday. Bottom line: it is NOT the same experience to read the *Times* on the Internet.

The decline in newspaper sales is not the only trouble. A steep decline in the number of classified ads is also slamming newspapers with a double whammy. Rupert Murdoch, who owns a bunch of newspapers, including the *New York Post* and the London *Sunday Times*, once described classified ad revenues as a "river of gold." Now, even Murdoch is predicting the demise of classified ads.

"I don't know anybody under 30 who has ever looked at a classified advertisement in a newspaper," Murdoch says today. He also thinks that their editorial strategies are to blame for the financial problems of U.S. newspapers. "Outside New York," he says, "it's all monopoly newspapers.

Some have good work in them, but it tends to be overwritten, boring, and elitist, not a reflection of the general mood in the public." Their circulations will continue to fall. But Murdoch also criticized recent attempts to boost circulation in the UK by the giving away of a DVD with each copy, a practice he thinks must stop. "I personally hate this DVD craze," he said.

Keep in mind that *you don't get results by solving problems but by exploiting opportunities*. Whether it's a DVD, café machine, or radio that is given to new subscribers, the incentive is beside the point. It will only create a demand for more and better incentives and does not address the problem of the drought in readers. Newspaper publishers are acting similar to the camera makers who overslept during the change from analog to digital. Giving away film or a photo album with each camera sold would not have solved their problem.

At the beginning of this section, I was careful to speak of the "slow death of the newspaper culture." The operative words here are *slow* and *culture*. I did not say, "the death of newspapers," though many newspapers have died and many are dying. What we are witnessing is the slow death of the newspaper culture, which had to do with the importance of newspapers in our lives. That aspect of our lives is clearly receding and of lessening importance. And as newspapers become less important in our lives, what happens with the newspaper enterprise is less interesting.

2. ADVERTISING—BACK TO "A PICTURE IS WORTH THOUSANDS OF WORDS"

Pompeii had its advertising blackboards, the Middle Ages its carnival barkers, America its Sears catalog. In advertising, nothing sums up what is happening quicker than that old cliché, "a picture is worth a thousand words." It is supposedly an adaptation of a Chinese saying: "One seeing is worth a thousand tellings." Either way, advertising is rushing away from copy to "visual narrative."

"No one buys a car off the page anymore," says London ad executive Rooney Carruthers. "They buy it because of image."

In contrast to the soft and friendly pictures with which Ikea visualizes what it stands for, Luciano Benetton's Italian fashion company chose a different path: advertising "not only a means of communication, but an expression of our time." In the 1980s, Benetton's campaigns shook up the world by using increasingly provocative pictures of our world. Much of the credit goes to Olivero Toscani, a radical agitator of throwing the truth right into your face.

In 1984, he started to exclusively head the company's advertising campaigns. He was a forerunner who shattered every conventional idea regarding the role of advertising. In his early works for Benetton, Toscani presented youthful images from culturally diverse nations. The varying colors of the Benetton collection linked with the diverse "colors" of its worldwide customers. With young people engaged in a variety of playful acts, Toscani presented a theme of racial harmony and world peace. It became the inspirational trademark still used today: "United Colors of Benetton."

This soft attitude shifted to the mindset that ads should draw attention to social problems and thereby generate discussions. Benetton's "Black and White" poster, showing a black mother breast-feeding a white infant, was designed to promote racial equality. Despite being Benetton's most awarded image, it sparked unprecedented controversy.

"La Pieta" was the most powerful image in the campaign; it posed David Kirby, a young man who was dying of AIDS, in his father's arms. Toscani named it "La Pieta," referring to Michelangelo's sculpture of the Madonna holding the dead body of Christ in her arms. The horrific reality and the pervasiveness of the disease indelibly burned these pictures into people's heads, when rushing to work, heading for shopping, having a good time, or being troubled by their own problems with no way to turn away.

Toscani was increasingly dealing with shocking images we see in the media and in the news all the time. But once put on a billboard, those pictures became much stronger and quite disturbing. People had to pause and think about their position.

Toscani, who headed Benetton's campaigns until the year 2000, explained: "Everything we do is about impulse, about guts. That's what built Benetton." It was pictures speaking across all cultural boundaries, raising social awareness by presenting powerful human and universal themes and making the larger point that communicating globally can be better accomplished visually rather than through language.

"I am not here to sell pullovers, but to promote an image [that sells pullovers]," said Toscani. Benetton said: "The purpose of advertising is not to sell more [I dare to doubt that]. It's to do with institutional publicity, whose aim is to communicate the company's values. We need to convey a single strong image, which can be shared anywhere in the world."

MULTIPLE CHOICE

The growing acceptance of alternatives to conventional print ads is one of the most significant changes in marketing in decades and a reflection of a further shift from word to visual—or to fewer words.

New media possibilities are exploding. Multiplying "toilet" ads allow companies to target gender with 100 percent accuracy (well, close enough). Bluntly a face cream ad on a women's bathroom mirror said: "It's not the lighting." Ads are everywhere: on napkins, cookies, bank cards, telephone cards, museum tickets, handrails on escalators, stickers on apples or bananas, the gigantic stems of windmills, Jetways to planes, in school textbooks, on the backside of plastic hotel room keys, and the backside of the advice in fortune cookies (sure good fortune for 17 million cookies produced by Wonton Food, Inc., in New York), painted on the bottom of beer glasses—and in children's names. Someone yelling "Budweiser" might not be desperate for a beer, but calling for his boy. In a recent poll in the U.S., 49 percent of people responding said they would at least consider naming their child "Coke" or "Kraft" or something similar if the company put up half a million dollars and 21 percent said they would accept an offer—if the price was right.

Are you thinking sports? Forget uniforms and Formula One cars. Skin is in! Volleyballers got it started and now some women tennis stars have offered to have temporary tattoos on their bodies. On the contracting muscles of a boxer's back, a casino's name might catch your eye. Recently Bernard Hopkins got a reported $100,000 to have the name of an online casino "GoldenPalace.com" inked on his back for a title fight.

Small potatoes compared to "product placement" in television shows and films ("Hated the movie; loved the product placement"). In the European Union, busy Brussels is already crafting rules regarding what is appropriate—or not—in media accepting product placement.

The value of product placement deals in films and television in the U.S. grew from $174 million in 1974 to $3.5 billion in 2004, according to PQ media, a market research firm. In five years, it could reach $7 billion, says PQ.

"Virtual product placement" is a process that uses computer graphics and digital editing to put products into television programs *after* they are taped. Multiple deals for multiple formats are available to advertisers: placing one kind of cookies in a first-run movie, another kind in the movie's DVD release, and a third kind of cookie in the portable video player version.

"Anything you can package, we can do. We could do pharmaceuticals, shampoos, takeout food, bags from Target," says David Brenner, president of Marathon Ventures in Wakarusa, Indiana, a big player in virtual product placement.

Advertising continues to become part of the entertainment business, and the biggest question mark is what role the Internet will play—besides big and important. One thing is clear: Advertising as a product of the twentieth century mirrored the mass production of a very successful mass production industrial era. That model is yesterday.

3. UPSCALE DESIGN FOR COMMON GOODS

The increasingly fierce competition of a global marketplace raises many new questions. The most urgent one is: As almost everyone has access to the same technology, what will make your company stand out? Once the novelty of a technology has worn off, the differentiation is through "high touch, and many are very well aware of it."

Ben Evans, director of the London Design Festival, said: "Design and creativity are one of the key competitive advantages companies in developed economies can have. In the future it will probably be the only one they have left." Robert Hayes, professor emeritus at Harvard Business School, blows the same horn: "Fifteen years ago companies competed on price. Now it's quality. Tomorrow it's design."

Scott Morrison summarizes in the *Financial Times* that the aim today has to be to create products and services that "look sharp, function intuitively and wake some sort of positive emotional response from the consumer."

Nicolas Hayek, chairman of the Swatch Group, put the theoretical insight into a practical context: "If you can combine powerful technology with the fantasy of a six-year-old, you can create miracles." And he did.

Not far from Zurich, his manufacturing buildings are scattered across a manicured landscape, enclosed by mighty mountains. His victory story started there in the grim mid–1970s. The Swiss watch industry was facing the threat of the Japanese, who had captured the market with the introduction of the Digital Delirium, a watch with the thinnest wristband and the least number of components. Instead of trying to solve the problem of being outperformed by going the traditional Swiss watch way of handmade, high-end watches, Hayek saw the opportunity in the new technologies and embraced them. He reengineered the structure of a conventional watch, reducing the number of parts from 157 to 50 and ended up beating the Japanese with a watch having a thickness of only 1.89 millimeters. The inexpensive watches were built in a virtually automatic construction process with only a handful of people in control. But despite succeeding on the technology side, there was still a bumpy road ahead.

The watch didn't sell. Hayek realized that there was no joy in a technological wonder with no life in it, no poetry. So he combined his high-tech product with high-touch color and flamboyant design. And then he trumped it all with a genius idea in marketing. In the Swiss world of high-end watches, Swatch created scarcity at the *low end*, limiting the number of copies made of each model. Now about 200 models are released each year, some of them designed by famous artists such as Keith Haring, Paul Bury, Sam Francis, and many more.

In the "message of the Chairman," Nicolas Hayek wears *the ecology of technology* on his sleeve: "The Swatch Group has a very special emotional culture. We produce beauty, sensuality, emotionality in watches—and we also produce high-tech on your wrists. Both emotionality or poetry and high-tech are part of what we feel towards our customers. We love them genuinely. We want them to be happy—we want YOU to be happy."

The company's day-to-day business might not be all that altruistic, but Swatch Watch has learned to listen to the heartbeat of our time. In a high-tech world, starving of imagination, high-touch product design is now as essential as the product itself. Design adds soul. Presentation distinguishes. Design is your stamp on your business.

A UNIVERSAL DESIGN LANGUAGE?

Design has emerged as one of the world's most powerful forces, a force with many faces. There is much talk about adapting product design to various markets around the world, translating it across different cultures. One emphasis at the moment is to design "platforms" or "architectures" around which products—especially automobiles—can be tailored to specific markets. But perhaps the problem is with the initial design itself. While U.S. automakers put their bet on different cars for different markets, Mercedes sells the same cars everywhere: "Only a Mercedes is a Mercedes."

Some products do speak a universal language of design that is received by all cultures.

THE DESIGN FLAG IS RUNNING UP
THE ENTERTAINMENT FLAGPOLE

Having hit the nerve of time and desires, Apple's iPod has swept around the globe like a music wave. Apple's iPod has become the world's most successful global generic product. There was "no need for regional adaptation or consideration for local languages," said Dylan Jones, editor of *GQ*. "One of the things it's easy to forget about Mr. Jobs's brilliant creation is that it is fundamentally international, and he has produced a machine that is truly pan-global."

Sometimes great design does not carry the day—even for Apple. "True, it looks like it belongs in the Museum of Modern Art," read an advertisement for Apple's supercomputer, the G4 Cube, a few years ago. Beautiful as it was, the Cube did not sell. Steve Jobs now says it was "not a failure of design. It was a failure of concept. We thought they would rather have something small on the desk than expendability, and we were wrong. It was a wrong concept—fabulously implemented." P.S. The Cube did make it to the Museum of Modern Art's design collection, while Nokia's simple slogan, "Connecting People," reached the market.

Nokia's slogan could not be simpler: "Connecting people." The basic insight is that as with other inventions, we have to understand that cell phones are not a technological phenomenon, they are a social phenomenon. It is connecting people; it is emotions. Think of Mindset #11 "Don't forget the ecology of technology": Once everyone has a cell phone, the relationship between consumers and their cell phones becomes paramount.

Nokia's chief designer, Frank Nuovo, remains amazed that there was no design competition for such a long time. "It was almost ten years before we saw anyone aggressively looking at industrial-designed mobile phones. So then they did turn up the heat and produce some decent designs. That's great—it makes it more of a challenge."

Nuovo, who is on the road 25 to 30 percent of his time, connects Nokia's ability "to cross cultural boundaries with handset designs that appeal across the globe" with the international makeup of its designers. He says he is particularly proud that 30 nationalities are represented in his design group, drawing inspiration from around the world and across different demographic groups and disciplines to create products that can continuously succeed in all markets.

As does Vertu, Nokia's luxury line of mobile phones, which seem to be treated like luxury watches, made with valuable materials like sapphire, leather, stainless steel, and precious metals, start at a price of $3,850. The standard kit for the Vertu MMII Stainless, not the high-end model, costs $7,200. You want to purchase the keypad you like with it? With golden keys? It will be an extra $1,200. There are no advanced features; it is just a magnificent body and expensive materials, nothing else. But it feeds the ego.

When I was joining a meeting in Santa Fe, Jim Taylor, author and Harvard professor of marketing, demonstrated the power of visual temptation. He pulled two ballpoint pens out of his pocket, a Montblanc Masterpiece and a no-name pen, jumped down from the podium, and asked someone in the audience to draw a few lines which each of them. Of course, there was no visible difference in the lines. "Now," he asked, waving the sheet of paper in front of our noses, "why would someone be willing to pay $1,500 for a pen that can do nothing more than the one for one dollar?"

Because after several years of hard work on a book, signing it with a one-dollar pen would be like eating Beluga on a plastic plate, I would say.

Ikea, Swatch, Mont Blanc, Apple, and Nokia are just some of the many companies that sell worldwide, creating a design that does not need to be adapted to different cultures. Unlike some products, where design overkills function, these companies seek functionality first. "Form follows function has been with us for a long time," as Nuovo puts it.

Asians, Europeans, and Americans are all mind and soul, reason and emotion. A product connects when it is in balance with both feeling and function.

4. ARCHITECTURE AS VISUAL ART

Have you noticed the number of celebrity architects who have become involved in store design? For decades, big-name architects turned up their noses at the thought of designing a mere dress store or a vulgar mall.

Today, architecture as visual art is more accepted than ever before. Big name international architects—Frank Gehry, Rem Koolhaas, Santiago Calatrava, Daniel Libeskind, Norman Foster, Richard Meier, Zaha

Hadid—are at least on a popular par with the celebrated artists working today: Gerhard Richter, Damien Hirst, Cy Twombly, David Hockney, Richard Serra, Robert Rauschenberg.

In 2004, the ninth Venice Architecture Biennale, the most important architecture exhibition in the world, was named "Metamorph." The Bienniale director, Kurt Forster, chose the name because it highlights the changing face of architecture, which "has taken on such breadth and depth as to suggest the advent of a new era." The tenth International Architecture Exhibition will be held in Venice from September 10 to November 19, 2006, with the theme "Cities, Architecture, and Society."

We'd better be aware of it, for even the U.S. federal government is getting into the act. Richard Meier's splendid U.S. Courthouse and Federal Building on Long Island is an outstanding example of a government's seeking and hiring the best architectural talent—versus the old days when the work went to architects with the best political connections. A milestone in this change in mindset about architecture and its commercial value was of course Frank Gehry's Guggenheim Museum in Bilbao.

Leading the Parade

A few years ago, on a sunny spring morning, we flew from Madrid to Bilbao, located in the north of Spain in the Basque region. The only reason was to see the Guggenheim, the ultimate project in the small city's redevelopment, which included the reconstruction of the city's airport and starting the first line of a new subway system. Also, supervised by Sir Norman Foster, the Uribitarte Footbridge over the Nervion River was built by Santiago Calatrava, and the waterfront project, a cultural and business area with two-thirds parks and open spaces, designed by Cesar Pelli.

A driver took us from the busy little airport to the town. He told us how much his business had improved and how the Basque region had come to life, and how making a living is much easier for everybody. When the building came in sight, its titanium waves were reflecting the bright sunlight in gold and beige, its utterly dazzling visual forms fitting into its ordinary neighborhood. With all the hype, I had been ready to be a tad disappointed. But the power and grace of the building made me understand why this iconic Guggenheim, which from the distance looked like the windings of a gigantic ribbon, had turned the region into a place of pilgrimage, setting a new level of architectural visual beauty.

This miracle was based on open minds and the belief that great architecture works anywhere. The mindset behind it was *seeking an opportunity and not solving the problems* of a declining population and fading-away industry. It was high risk and a great success. Bilbao was leading the parade of selling contemporary great architecture as a destination. The price of titanium, which is the essential material of the building, had come down with the cessation of atomic bomb production, to a level where it could be used as a building material. Without the computer, the dynamics of the design could not have been calculated and translated into the specifications for the actual building of the museum.

None of that crossed my mind as I stood there enjoying. I became a pilgrim who would return for another religious experience, a pure and exuberant visual pleasure that changed my ideas of what architecture was and could be.

More than 70 years ago, the New York Chrysler Building by William van Alen, built in 1930, and later the 1954 Seagram Building, by Mies van der Rohe, set the standards for high-rise buildings of elegance, simplicity, and clarity. Well-designed, distinctive architecture has become a great branding concept for museums and companies. Because of iconic architects, "architecture as art" is probably more widely accepted today than ever before. Swiss Re's building in London is now known as the "Erotic Gherkin," looking something like a brilliant glass pinecone. It has won many prizes and made the insurance group a much more familiar brand to millions of Londoners and visitors. What a contrast to buildings that are distinguished only by company names in giant letters.

A New Search for Identity Through Design

When New York's Museum of Modern Art (MoMA) began to organize a large show about architecture and public places around the world, one of the first stops was Spain. MoMA's chief curator of architecture and design, Terence Riley, became so impressed with the architecture performance and excitement about the changing man-made landscape in Spain that he decided to change the show from global architecture to a show about Spain as a global leader in architectural design. It all had started with Frank Gehry's Guggenheim Museum opening in 1997. The turnaround it brought about, said Riley, prompted other cities in Spain to ask, "Why not us?"

"Local mayors have been emboldened, risk has been incentivized," he said. "Post-Gehry, there is this whole attempt to make the country more open to cultural tourism." There is, without doubt, an architectural renaissance taking place in Spain, presenting a visual showcase for the rest of the world. Frank Gehry's museum is an example of how *the future of culture is embedded in the present.*

After decades of fascist rule, the democratic awakening has been powering Spain. And before the Franco period, as Riley points out, "in the eighteenth and nineteenth century, [Spain] clung to this Counter-Reformation definition of itself. It became the antimodern state, autocratic and orthodox." Today's activity "is a serious attempt by these cities to address the deficit between them and other countries of Europe." Spain today is dramatically outpacing other parts of Europe in the extent and daring of its new architecture. As a measure of all the architectural activity in Spain, construction is now the largest economic sector in the country, larger even than tourism, a remarkable development considering that Spain has for years been ranked either the number one or two tourist destination in the world (alternating with France).

While such celebrity architects as Zaha Hadid (three projects), Jacques Herzog and Pierre de Meuron (seven projects), Jean Nouvel (five projects), plus Rem Koolhaas and Frank Gehry (a new hotel in La Rioja), and Richard Rogers are being presented, of the 53 projects in the Museum of Modern Art show, 70 percent were designed by local architects. Spain now holds architectural competitions for all new public buildings, some of the competitions only for people younger than 40. The MoMA show opened in New York in February 2006, a showcase of the visual world.

THE HIGH-PROFILE TRAVEL CLUB

The shift to a visual world, where design stimulation can be the USP of a region or product, has opened doors into new business opportunities. One of those doors led to designer hotels. Good for us. My wife and I travel a lot and we are not aloof to visual enchantment. A couple of years ago, as we were on our way to Oxford, we decided to make a stop in London, it was no question where to stay: in the Sanderson, Ian Schrager's highly praised Philippe Starck–designed hotel, north of Soho. It had opened its doors into "a magical stage for the world's high-profile travellers."

Expectations were flying high when our car stopped at the building, a late-1950s company headquarters, now the Sanderson disguising its internal surprises. "Sanderson is all about mental game," Starck had said. "It is the most culturally complete space we've done. It's difficult to understand. It's a hotel of illusion—optical illusion. Poetic tricks. You don't know what you see. Where are the walls? What is the style?"

How to characterize our feelings when the door opened? Well, it was about like opening Aunt Mary's present on Christmas day. What to say?

"You have to understand that guests become actors in these spaces," Anda Andrei, president of design at Ian Schrager Hotels, had revealed in her *Vanity Fair* interview. "People do not simply enter the lobby. They may instead ask themselves, do I look cool right now as I am walking through the lobby?"

The jungle-drum stools in front of the reception desk were pretty hard to sit on. We might have slipped into a less stylish posture while sitting there waiting for our room keys, but that all changed again when the outrageously friendly staff guided us to the elevator that took us to our floor. We had Starck's words in mind: "In this elevator you fall into the universe. You fly. There is a promenade of culture."

We left the universe on the fifth floor and walked over heavy carpets to our room. I opened the door. Minimalist. White. Light. Small—why had we not booked a suite? But that, of course, was our fault.

"It's the most ethereal space I've ever seen," Andrei had said. "It's a bit feminine. It's very sexy. If I were staying here, I wouldn't leave the bedroom."

My wife turned to me: "Don't you think this looks like one of those rooms where you wait to go into surgery?" she said, not amused and demolishing any attempt to really distinguish us for the high-profile traveler club. Fully aware that we simply had not made it into "the sophisticated fog, the mental cloud, where every sign opens a door into the brain" (Philippe Starck again), we moved to an old-fashioned hotel nearby the next morning.

Overdesigning is somehow like the embracing of new technology without thinking of its ecology. What is enhanced? What is diminished? What is replaced? Less is more, but leave something! Designer hotels find their clientele as long as it is hip to stay there, but the novelty wears off and the basic needs remain. As diversified as the definition of taste is, self-serving design is not the answer.

Last year, the otherwise sensible *Financial Times* said in an editorial that conspicuous consumption is out and self-fulfillment is in, through "luxury experiences," noting that Armani and Bulgari have moved into the luxury hotel business. "Luxury experiences" may become more in the mix, but will not replace other forms of conspicuous consumption.

5. FASHION, ARCHITECTURE, AND ART

In a new book on art and fashion, Alex Coles remembers legendary art critic Clement Greenberg, "who said something like, 'art has always had an umbilical cord of gold attaching it to the money people'—but that cord also allowed it a critical distance. By placing art in the actual context of the money world, that distance is completely collapsed." Coles was pointing to "a major shift" in the *degree* to which fashion companies are embracing art.

No wonder that in a world oversupplied with luxury goods, prestige brands are looking for new ways to entice us. The luxury goods markets are saturated—oversupplied—and there is too little demand. *Solving a problem or seeking new opportunities?* What to do? Or, as Karl Lagerfeld put it: "How do you keep desire alive?"

Exploiting a symbiosis of art and fashion is the answer. High-end brands such as Dom Pérignon and Chanel are turning to creative five-minute films to generate a fantasy, serious art context for their products. Fashion houses are moving beyond standard commercials by turning to star directors and supermodels. Australian filmmaker Baz Luhrmann perfectly handles the power of storytelling for Chanel, directing Nicole Kidman as the woman of one's dream through euphuistic scenes. Armani turned to Michelle Pfeiffer, who called him her soul mate, and invites us to see the world through his sunglasses. Uma Thurman is TAG Heuer's global brand ambassador, and George Clooney's charm melts into Nespresso's taste. LVMH (Möet Hennessy Louis Vuitton) general director Jean Marc LeCave says, "These days we need names that are recognized the world over, from Tokyo and Shanghai, New York and Paris."

Karl Lagerfeld took the photographs for Dom Pérignon's DP98 campaign and merged desirable images to a new picture: the tingle of champagne, the beauty of a woman, the sense of a lifestyle, the attraction of the unattainable. Karl Lagerfeld's visuals transmit a fairy-tale lifestyle for those who dream of it.

Andy Warhol once said that "fashion isn't what you wear someplace anymore; it is the whole reason for going."

THE THIRD IN THE ALLIANCE

Never was the golden cord between art and money so refined and well fed as in the long history of the aristocrat world and its patronage. Kings and queens, popes and potentates were encircled by art and poetry, music, and grand architecture. The third in the alliance, Louis XIV, le "Roi Solei," attached his name forever with the most prominent imperial castle, Versailles. The sovereigns of our time, the moneyed noblesse of fashion and commerce, are returning to this culture again, braiding together the three elements of today's umbilical cord: art, fashion, and architecture.

When you walk through what is called Paris's golden triangle—avenue Montaigne, avenue George-V, and rue François-1er—the big names in fashion are lined up like sparrows on a wire: Dior, Chanel, Rochas, Hermès, Cartier, Prada, Issey Miyake, Yohji Yamamoto, and so on.

Within all this lavish luxury, a seven-story building stands out: Louis Vuitton's Art Deco luxury flagship, the world's largest building for a single brand. On the top floor of the building Vuitton has created a serious art gallery, Espace Cultural Louis Vuitton. A spacious 4,300 square feet with a 16-foot rotunda are waiting for the public; a terrace running all along the outside offers a stunning panoramic view of the city. Though visitors can reach the gallery from inside the store, it has its own entrance at 58 rue de Bassano. An elevator, designed by the Danish artist Olafur Eliasson, is an experience on its own. In a soundproof, totally dark space, "the visitor is invited to experience his or her own inner voyage of self-discovery" in a 20-second ascent into the exhibition space. Not to be too demanding, but if your inner space is of any substance, 20 seconds is a pretty short ride.

In January 2006, when we visited the Louis Vuitton building, the opening elevator door catapulted us from our inner space directly into the nudity of Vanessa Beecroft's "Alphabet Concept." Beecroft has created this completely new set of artwork for her collaboration with Vuitton. The work is composed of 13 photographs in which the LV logo and the letters of the label's name are spelled out in arrangements of nude female bodies. "Nudity becomes its own kind of outfit, not unlike the old emperor's clothes fairy tale," she says.

Beecroft, an Italian-born artist known for her images of good-looking female models taking off their clothes and standing around in galleries and museums, had already staged the memorable near-nude performance in October 2005 for Louis Vuitton's huge opening party celebrating the Champs-Elysées store in the nearby Petite Palais.

The LV gallery will show five exhibitions a year, concentrating on four themes—heritage, art, fashion, and travel—that have inspired the Vuitton brand throughout its 150-year-old history. The CEO of Louis Vuitton, Yves Carcelle, says that within his company, "there was a real consensus that a luxury house could be a vehicle to put art, and especially modern art, in contact with the public. Museums can be intimidating—you feel you shouldn't talk too much in a museum. Here you can, and that creates a different relationship with art." The store itself is full of avant-garde commissioned art by established contemporary artists. Is there a concern in the art world that these artists are selling out? Well, maybe. Vanessa Beecroft sums up the conflict:

> The bourgeoisie like to use the intellectuals to clean up their soul;
> they think progressive, cutting-edge work is good for their image. So

did they use me? Sure, and fairly cheaply—they didn't pay that much, just like a museum. But in some ways, I like to be used. I felt threatened in the store—it's too much, too many shoes, too many bags, too much with the art—but I like to feel threatened; it's like being in a war zone.

Peter Marino, the architect who designed the store's interior, asked: "When the great Renaissance painters did the Medici chapels, were they selling out?" and answered, "No, that's what lasts. We don't have popes commissioning art anymore; we have big corporations."

The umbilical golden cord still holds.

Nearly a century ago, Mario Prada started the Prada label in Italy. Since 1978, Miuccia, Prada's granddaughter, is running the company. Under her leadership, the Prada look has become a leader in innovation in fabric and design. Together with Patricio Bertelli, she founded Fondazione Prada in 1997, an important art gallery in Italy, and did the restructuring of the contemporary art foundation PradaMilanArte, hosting two shows of international artists each year.

A different kind of show is what you get when you walk into Rem Koolhaas's Prada store in New York's SoHo. It cost a cool $40 million for 23,000 square feet and sells practically nothing, but it gives the customer a visual pleasure to compliment the fashion on offer. "It really fits in the strategy with a luxury brand because you have to offer clients a unique luxury experience from A to Z," says Prada.

For Prada, these stores are important communication tools: "So the returns are not just measured on how many handbags we sell in these stores, but also on how such stores define our image." And in fact, when the store was opened in December 2001 it received bundles of publicity. "Think of this as a museum show on indefinite display," wrote Herbert Muschamp, the architecture critic for the *New York Times*. The opening event was enjoyed by celebrity guests including then New York City mayor Rudolph Giuliani—and thus it would seem the design did at least partially what Prada wanted.

And Prada is staying in the game. Since the Rem Koolhaas SoHo opening, it has opened two more flagship stores, one in Tokyo, designed by Herzog and de Meuron, and another Rem Koolhaas, plus Ole Scheeren, in Los Angeles's Beverly Hills. Prada has now spent more than $100 million on these flamboyant stores, and it looks as if Shanghai is next.

In this scene, Frank Gehry's contribution to the fashion world held what his Bilbao museum promised. Issey Miyake contracted "Bilbao fever," and the Tribeca store near Prada was the result. It is a breathtaking, delicate balance between shopping and in-store entertainment. Whether you come to buy or peer, you'll wander wide-eyed through this pure Gehry space, where a monstrous titanium tornado is swirling through a two-story, 3,000-square-foot boutique. Is high-end shopping entering a Bilbao phase, or are these just passing conceits and indulgences? Frank Gehry said, "I think Issey and I may be after the same thing in our work. We're both trying to express movement and play around with new materials that haven't been used before."

The design virus of high-end fashion shops has been trickling down to retail selling space below. With experiential shopping increasing at the high end and Internet shopping increasing from below, retail design has and will become more important.

6. MUSIC, VIDEO, AND FILM

Anytime, anywhere.

In 1957 it started with a plop. The coin disappeared into a slot. A row of buttons was eagerly waiting, raising their square bodies with numbers to be pushed. Click. Click. Black disks, lined up like soldiers, started to move in a crabwise parade, purring softly till a sudden halt. A half-round arm appeared and tightly embraced a disk, lifted it, bent, and put it down on a turning plate. A second arm appeared, long and thin, reaching out with a sharp pin at its vibrating end. Gently it moved down until the pin found footing in the endless groove of the evenly turning disk. *Palop, palop, palop*, it swung in the rhythm of turning until the music started to play—the jukebox.

Today's jukeboxes have shrunk, but what they offer has multiplied. Dell's jukebox, called DJ, lies in your hand like a deck of cards and plays for up to 16 hours—good for about 100 vinyls to choose from. But music alone doesn't do it anymore. It's visual, it's music videos, it's MTV.

MTV has contributed mightily to the shift to a global visual culture.

MTV started with almost exclusively broadcasting music videos at three minutes a pop, a new TV entry that was to change our visual culture and exert exceptional influence on experimentation and marketing ("I want my MTV").

"MTV's durability at the place where the fickle music business and the protean television trade interest intersect can be attributed to a singular mindset: its 24-year-long insistence that the channel itself is the star," wrote Kate Arthur in the *New York Times*. The talent can come and go, but MTV endures.

MTV Networks now has 112 channels around the world, including Comedy Central and Nickelodeon.

MTV can be held accountable for advancing the culture of design and shortening the attention span of viewers.

MTV has since moved to broadened programming and "videos have taken on an exciting if uncertain life of their own, far away from the mother ship that launched them," reported Jon Caramanica in the July 31, 2005 issue of the *New York Times*. The proliferation of videos on new screens has been as quiet as it is profound. To the global visual world, videos energized visual marketing, with Prince, Michael Jackson, and Madonna all reinventing themselves through video. The video, as Caramanica points out, is now one of the only forms of media that truly extends to all screens and devices.

The arrival of the video iPod joined two of the most profound innovations in pop music, which presages having the whole visual world in your hands. Attach your WebCam to the computer, create your own reality show, and get famous. The Internet is about to become the hot spot for wannabe and real stars. The thrust of all this, chaotic as it may be, is to get more images to more people faster, exploding a visual world. Video world premieres on Yahoo!, AOL, or YouTube, downloaded directly to your iPod. The avant-garde keeps moving.

Video Games as Art Form

Video games, which have collectively taken up much of the space and time television once took, are gaining attention as an art form. Americans now spend more money on video games than on movies. And according to University of Wisconsin Professor James Paul Gee, video gaming has become "the major cultural activity of the generation 30 or 35 and below, the way movies and literature were for earlier generations."

The history of media is that the new forms initially borrow the content of the previous medium. Radio at first broadcasted plays; television filmed radio shows and showed old movies. Video games tried movies but with disastrous results. It is widely known in the gaming industry that a game based on the movie E.T. was the biggest failure in game history. But "cinema envy" still affects the big companies in this $10 billion industry, says Eric Zimmerman of gameLab. The impulse to make people cry, he says, was a "misguided idea of what emotional depiction is." Zimmerman says that "games are by nature incredibly emotionally engaging. Games are dynamic, participatory systems. That's a level of storytelling that a film can't do."

And Professor Gee adds, "We haven't begun to scratch the surface of what games we could make without somebody shooting someone."

Storyteller Steven Spielberg, who is very involved in the gaming medium, says that the medium will come of age "when somebody confesses that they cried at level 17."

In the December 4, 2005, issue of the *New York Times,* in an excellent article called "The Gamer as Artiste," John Leland reported that museum exhibitions, academic conferences, and university curriculums have examined games as art. A 2004 conference at Stanford University called Story Engines looked at game playing as a way of creating narratives, "at a time when the audiences for established story vehicles like books, newspapers, movies and networks are in decline."

Who are today's storyteller successors? To anticipate the future, watch the score of the game. In the United States, video game sales in 2005 were more than $10 billion, with the two makers of most popular consoles, the Sony PlayStation 2 and Microsoft Xbox, just gearing up to introduce new consoles. The two top-selling games were *Grand Theft Auto: San Andreas* and *Halo 2,* each selling more than 5 million copies. Perhaps no one reading this has played the games, but for many young people around the world they are part of the passage into our visual world.

FILM

As confused as anyone about where our visual world is going, Hollywood was trying something new in spreading its images in January 2006. A film called *Bubble,* a murder mystery set in a small town in Ohio, was the motion picture industry's first "universal release," made available simultaneously in movie houses, on home video, and on pay television.

The Internet entrepreneur behind *Bubble,* Todd Wagner, sees this as the film industry's late entry into the digital age. "To me, it's pretty clear that times have changed," he said. "What is changing here is Hollywood's practice of telling customers where and when they can consume its products. If you did that in the music business, it would be the equivalent of saying, 'Well, if you like that song on the radio, you'll have to wait five months to buy it.'"

Underscoring the changing landscape, Rick Finkelstein, president and chief operating officer of Universal Pictures, says, "Eventually we're going to have home networks and people will be able to watch whatever they want, whenever they want on whatever device they want."

Again, *what is the score of the game*? Since 2003, U.S. box-office sales have been about $9.5 billion each year. Last year's DVD sales alone hit $15 billion, up 30 percent from the previous year. DVD releases grew from 1,500 in 1997 to more than 11,000 in 2005. And, for the record, Wal-Mart is the world's leading vendor of DVDs.

The next big move in the works: satellite digital distribution directly to the nearly 110,000 screens in movie theaters around the world. DVDs are also a big part of the global trend toward the decentralization of entertainment: from entertainment for the masses to massive entertainment for the individual. It's a whole new business. In fact, in 2003 alone, consumers spent $14.4 billion on movies for the home. That's *billions* of dollars more than they spent on tickets for the cinema.

This massive collecting of DVDs by American consumers is huge, and largely unanticipated. Yet while DVDs have been a wonderful addition to home entertainment, they lack the capacity to offer both high-definition content and broadband interactivity.

Finally, the world is increasingly embracing the new technology because of growing global affluence, an incredibly expanding global middle-class market for electronics and entertainment. This demand is so strong, it supersedes the new technology standards wars that are inevitable. In earlier times, governments decided on standards—despite the fact that there is no way any government can anticipate the needs and desires of millions of potential customers.

Today, the marketplace defines the standard, so in the larger context, legitimate decisions concerning standards can be made only by consumers. This is the overwhelming global trend. The convergence of technology and entertainment that this industry has been talking about for so long is a truly dynamic process, where each step instructs all succeeding ones.

In the final analysis, this process must fully engage the consumer before *any* new technology ultimately becomes a success. As the world moves increasingly closer to the age of high-definition media, the free market will decide what is compelling—and what is not.

7. THE CHANGING ROLE OF PHOTOGRAPHY

Do you know the land where the lemon-trees grow,
In darkened leaves golden oranges glow,
A soft wind blows from the pure blue sky,
Where the myrtle stands mute, and the laurel high?
Do you know it well? There,
There my love with you I shall be.
JOHANN WOLFGANG VON GOETHE, *1749–1832*

For centuries, the only way for many to explore the visual beauty of the world was through someone else's eyes, the brush and easel of a painter or the narrative of a daring traveler. Marco Polo painted pictures of an unknown China; Goethe introduced us to his Italy. Led by personal experience and emotions, those pictures of reality were not so real after all. Then photography came along and made it possible to look through a lens's veridical eye.

TV changed everything by bringing remote countries and cultures into everyone's living room. Photography survived it all, but its purpose started to change.

The camera is an instrument of detection. We photograph not only what we know, but also what we don't know.
LISETTE MODEL, *1901–1983*

Lisette Model was born in 1901 into a wealthy Viennese family. She developed a straightforward and direct approach to photography, shooting both physically and emotionally from the gut with her Rolleiflex, closely cropping images of her subjects for a direct and powerful impact. Her 1958 series of portraits of people on the fringe of society made her well known.

At that time, around 1957–1960, the young Diane Arbus studied photography at the New School in New York. Lisette Model was her teacher. Arbus had grown up in a rich and privileged environment on Central Park West, lacking nothing in terms of possessions. Her wealthy parents sent her to the best progressive schools in New York. Her approach to photography was similar to Lisette Model:

It's important to take bad pictures. It's the bad ones that have to do with what you've never done before. They can make you recognize something you hadn't seen in a way that will make you recognize it when you see it again.
DIANE ARBUS, 1923–1971

The Museum of Modern Art (MoMA) called her "one of the most original and influential artists in the 20th century." From March to May 2005 the museum ran the exhibition Diane Arbus Revelations.

In what is known as the classic photography market, a Diane Arbus print of "Identical Twins, Roselle, New Jersey" recently sold for $478,400 at a Sotheby's auction. Robert Mapplethorpe's "Calla Lily, 1986," sold for $242,700 at the same auction. Six-figure prices have become routine. In 1999 Man Ray's "Glass Tears" (one of six known prints) sold for $1.3 million, the highest price ever paid for a photograph (so far). Digital photography, with which you can outrageously manipulate images, has made the old, classic photographs seem more authentic and valuable. But now there is a big market for digital photographs. A huge, digitally manipulated 1997 photograph by Andreas Gursky recently sold for $613,000, a record for any contemporary photographer.

But photography has had a hard time being recognized by MoMA. When Alfred Barr set up MoMA in 1929, he made architecture and photography foundations of our visual arts culture, explicit participants in the museum. In 2004 the museum put on a fashion photography show and in doing so congratulated itself on presenting the first major fashion photography show. The self-congratulations rang a little hollow. The show, called Fashioning Fiction, turned out to be as boring and pretentious as installations and videos and certainly arrived a little late, considering that for the last couple of decades MoMA and other museums have presented "contemporary art" shows consisting of vapid installations and boring video pieces, all of questionable artistic merit. But it is a beginning, although someone said that if MoMA hadn't been temporarily located in Queens (while its Manhattan headquarters were being redesigned), the show never would have happened.

On February 14, 2006, at a Sotheby's auction in New York City, a 1904 photograph was sold for $2.92 million, the world record for any photograph for any period. It was "The Pond, Moonlight" by Edward Steichen, the first of the twentieth-century giants in American photography. Such a price underscores at least a subliminal understanding that we are living in a pictorial world.

8. THE DEMOCRATIZATION OF THE AMERICAN ART MUSEUM

In recent years the art museums in New York City have regularly outdrawn all the city's sports events put together. The number of U.S. museums is multiplying rapidly, and many are designed by celebrity architects. Blockbuster art exhibitions have been setting record attendances. Branches of New York's Guggenheim Museum are spreading around the world like mushrooms in the woods. The most spectacular of course is Gehry's Bilbao Guggenheim, which has been attracting more than 900,000 visitors each year; add to that branches in Venice, Las Vegas, and Berlin, with talks underway for branches in Rio de Janeiro, Hong Kong, St. Petersburg, and Guadalajara. The joke in art circles is that eventually, there will be more Guggenheims than McDonald's.

Contemporary art has become the chic world's most covetable commodity. As Emma Crichton-Miller wrote in the *Financial Times*:

> Contemporary art is neither just a product nor a private cultural pleasure; it has become a way of life. It was once the yacht, the sports car or racehorses that marked one out as smart, cool and successful. Now it is an art collection. Today it is a painting by Gerhard Richter, or a sculpture by Takashi Murakami—that shows you have arrived. Galleries are now at the epicenter of High Society.

The Top of the Pyramid

One of the exciting games in our visually oriented world is the market for fine art, and the score of the game is recorded by the auction houses that sell the paintings and sculptures. World-record prices are tumbling everywhere as the highest prices ever are being paid.

At the summer 2005 auction in London, Sotheby's sold Kees van Dongen's *Femme au grand chapeau*, a 1906 portrait of a prostitute, for $9.2 million. A landscape painted by Paul Signac in 1886 went for $6.6 million. Two world records in the old masters field were set on successive days at Sotheby's with the sale of two works by Canaletto, the Italian artist who painted those beautiful Venetian views of the Grand Canal. The first, painted in 1754, sold for an unbelievable $20 million. Then *The Grand Canal from Palazzo Balbi* sold for $32.6 million, beating the record set for a Canaletto the day before.

We have become accustomed to sky-high prices for Van Gogh and

Picasso paintings (Picasso's *Boy with a Pipe* sold for $104 million in 2004), but as the supply shrinks and demand from a world filling up with billionaires increases, so will prices for many artists. In June of 2006, the cosmetics magnate Ronald Lauder paid $135 million for the 1907 portrait of Adele Bloch-Bauer by the Austrian Gustav Klimt, the highest sum ever paid for a painting.

The President's Medallion

I confess an inordinate interest in art. I have been a collector of modern art for more than 30 years, and sometimes I say I write books in order to buy art, which is close to the truth. This devotion to art has gotten me in trouble more than once.

While I was working for the White House, President Johnson, who always wanted to announce something in his speeches, announced a new program called Presidential Scholars. Because of my interest in art, I was asked to select a sculptor to create a representation of the president for the medallion that was to be awarded to the scholars.

For several days, I called all the artists and dealers and art critics I knew for suggestions. The name that came up most often was Jacques Lipchitz, one of the great sculptors of the twentieth century, who was living in New York City. I called Lipchitz to ask him to sculpt the president. He immediately agreed. "It would be a great honor."

Jack Valenti (later head of the Motion Picture Association, and then close personal aide to the president) provided me with a good-sized stack of profiles—"only the right side"—of the president. I took them to Lipchitz, who had asked for permission to do an initial likeness in clay, before sculpturing the president in person, "to breathe life into it." A problem, I had thought, but word came back that the president would do it.

The Sunday before the chosen day, a Monday, I went to New York to get Lipchitz and installed him in the Hay-Adams hotel across Lafayette

Park from the White House. On Monday we waited in the outer office for several hours, only to be told at the end of the day that it would not be possible to see the president on that day. It became a daily ritual—pick up Lipchitz, walk to the White House, wait, trudge back to the hotel. With each day, Lipchitz, already in his 80s, got increasingly tired and discouraged.

Finally, on Friday at about 8:00 PM, we were told that the president was very busy but we could come in to watch the president while he signed correspondence. Once with the president, Lipchitz came to life, furiously adding small pieces of clay to his 12-inch model. The president hardly noticed.

On the walk back to Hay-Adams, Lipchitz was very excited about having met the president. "He is such a big powerful man." Lipchitz, a bear of a man himself, said, "He ... he ... has balls!" In his struggling English, how could he say it? "Maybe three," he sputtered.

The following week I went to New York to get the finished model. Lipchitz's profile of the president was a very powerful, "three balls" rendering. I unveiled it in Walter Jenkins's office in the West Wing (Jenkins was the president's chief-of-staff at the time). Several of Jenkins's staff joined us. The rendering was greeted with stunned silence, finally broken by one of Jenkins's assistants: "When the president sees this, the Lipchitz is really going to hit the fan."

The president hated it. Hated it. But the president's wife, Lady Bird Johnson, saved the day. She, who knew about such things, told the president that rejecting Lipchitz's work would result in a big scandal in the arts community, an important constituency for the president. He didn't understand the art, but he understood the politics.

Exploiting the Opportunities of the Increase in Visual Demand

The confluence of globalization and the emerging dominance of the visual arts have led to a new field of freelance international art curators who are hired to curate the exploding number of art exhibitions. Curators were usually tied to the museums they worked for, but now the best of them have cut themselves loose to serve, among other shows, the more than 200 art biennials taking place continuously all over the world today, including those in Johannesburg, Istanbul, Cairo, Kwangju (Korea), Taipei, Yokohama, and Shanghai (also in China: Guangzhou and Shenzhen).

The esteemed art magazine *Art+Auction* credits a Swiss curator, Harald Szeemann, who became a freelance curator-director in the 1980s, with inventing the business model for putting together large-scale art exhibits. Some years ago he cast himself as an independent, international impresario. It was Szeemann who revitalized the Documenta in Kassel, Germany, and who inaugurated the Aperto at the Venice Biennale. The Venice Biennale was the first biennial, founded in 1893 as a world exposition of art, beginning its exhibitions in 1896. Szeemann introduced the Aperto as part of this biennial in 1980. The Aperto, as *Art+Auction* reports, was the encyclopedic show "at the heart of the biennial that was intended to circumvent the limitations of the national pavilions [which followed the model of the nineteenth-century World's Fair expositions] and emphasize younger, undiscovered artists." Today most of the itinerant art connoisseurs at one time or another attend the great-grandmother of all biennials, the Venice Biennale (I have attended all but two of them beginning in the 1980s), while the grandchildren of the first biennial are scattered throughout our visual world.

All art cannot be presented, no matter how many biennials there are. That opens opportunities to people whom we can trust to select and arrange art for presentations. The role of art curators is in tune with the times, and I expect it to be a greatly expanding field. The Internet will keep word editors busy forever.

The number of sponsored museum shows such as Armani and The Art of the Motorcycle at the Guggenheim Museum and Chanel at the Metropolitan Museum of Art in New York is increasing. But when New

York's MoMA inaugurated the temporary art galleries on the top floor of its new building in February 2005 with a selection of 64 works from the UBS (Union Bank of Switzerland) art collection, *New York Times* critic Roberta Smith complained that it was "an appalling paean to a corporate sponsor's blue-chip collection." Such complaints are rare these days, and, in fact, 40 of the 64 works in the UBS show were gifts to the museum. Corporations are shifting their focus from acquiring art to supporting art. "For probably a decade now," says Marc Porter, president of Christie's Americas, "American corporations have not been especially active in the market. They have funneled their art money into community cultural programs rather than buying objects themselves."

Porter points to Chase, which has a collection of more than 20,000 works of art, partly because of its merger with JPMorgan in 2001, which also had a substantial collection, and is now supporting museum shows and exhibitions of its own collection.

UBS has one of the world's best corporate collections and is now focusing on museum shows. Reputation rather than investment is now driving UBS's role in the visual arts world.

But some companies are still buying art, notably Deutsche Bank, now the papa bear of corporate art collectors with more than 50,000 pieces. Friedhelm Hutte, the cocurator of the collection, says that neither investing nor marketing motivates the bank's art purchases. He considers art "an alternative, emotional and spiritual dividend" and believes the company's support of young artists is a "social responsibility."

In America, corporations provide a lot of support for community arts projects, but in Europe corporate support is still very thin or nonexistent, making Deutsche Bank all the more outstanding. "It started 78,000 e-mails, 36,000 letters, five years, and 1,455 works of art ago," said Michael Klein five years after he was chosen to became the art curator of Microsoft. Even for Microsoft life is not all 010101.

Klein passed an interview and "did interview really well," one of the ex-members of the committee wistfully said when it turned out that they were not always in harmony.

"I showed them a print I thought we should acquire by Alex Katz," said Klein. "Great piece, and I got a deal on it. Somebody raised his hand and said, 'You're not going to shove your aesthetic down our throats.' That was it. Meeting over. I told Microsoft it was them or me. Microsoft wanted me to work there; it needed to know I wasn't going to work with a committee."

He stayed. The committee left. At 52, Klein is emerging nationwide as an important curator of a corporate art collection. He's good, but that's not the only reason his star is rising. At a time when many other corporate collections are shrinking or disappearing altogether, Microsoft's is expanding. "Of course it's expanding," said Klein. "Since I've been here, there are 12 new buildings on this campus. My job is to make them interesting through art."

While the preceding eight forces are leading the push to the visual world as the drum majors, many others are just joining the parade.

COLORIZING THE WORLD

THE POLITICAL "COLOR REVOLUTIONS"

In the spring of 2005 the *New York Times* ran a story about political uprisings in various parts of the world. The banner headline read: "The Revolution Will Be Colorized."

Part of my visual-world thesis is that the whole world is being colorized. By the way—a touch of irony here—the *New York Times* itself made the revolutionary step to be colorized only a few years ago when for the first time it began to use color pictures and photographs on its news pages. Now it does a better job than anyone with its color photographs.

Color associated with political activism was certainly given a huge boost by all that orange we saw in the dramatic uprising in Ukraine. "Those thousands of people in Ukraine wearing orange didn't even have to open their mouths," said Leatrice Eiseman, director of the Pantone Color Institute. "You knew what they stood for."

Since the Orange Revolution in Ukraine, we have been witnessing new attention to a color standing for a cause. Years ago we rather matter-of-factly associated red with Communists and black with Fascists, and for many years before that color had been used in flags and banners. Now political movements are drawing on a spectrum of colors to associate with their cause. The first modern such use I remember was the color yellow—used at rallies to support Corazon Aquino in the Philippines. In Iraq, purple, the color of the ink voters dipped their fingers in, became the color to signify that they had voted in elections. In Iran, pink is the color of dissent; in Belarus, blue; in Kyrgyzstan, yellow and pink. We saw the Rose Revolution in Georgia. In the United States, we now have red states and blue states.

Using color as a weapon is neatly summarized by Karen Beckwith, a political science professor at the College of Wooster in Ohio:

> How does the state respond to it? It's very hard to defeat. You can't go around making people take off their clothes. Also, the state can't tell who's organizing it. And it shows incredible solidarity. You know that you're not alone. You don't even need to carry a sign. The person himself or herself is the protest.

One measure of the effectiveness of political color is that the Chinese political functionaries are so freaked out by all the uprisings and their own vulnerability that they regularly speak out against the "color revolutions." When Beijing stopped plans to let foreign newspapers print in China, Shi Zongyuan, the top press regulator, put it plainly: "When I think of the 'color revolution,' I feel afraid." In Beijing, many authorities see the "color revolution" as a possible threat to the party's dominance. In November 2005, the official *People's Daily* ran articles condemning the U.S. media for shaking the "ideological mindsets and cultural foundations" of other countries by exporting U.S.–style values of freedom and democracy.

With President George W. Bush calling for the global promoting of democracy and "color revolutions" bringing down governments in Georgia, Ukraine, and Kyrgyzstan, Chinese Party officials are nervous, but most Chinese people are too busy exploring new opportunities for making money to notice.

COLOR MESSAGE BRACELETS

In May 2004 Nike produced 5 million "Livestrong" bracelets to honor Lance Armstrong as a cancer survivor in his quest for a sixth consecutive Tour de France victory. They were sold for $1 in retail outlets around the country, the money going to the foundation Armstrong set up to help fight cancer. Nike says that more than 50 million have been sold.

Armstrong won the sixth Tour and the next one, his seventh, in 2005. It was Armstrong's yellow bracelet that launched the color ID craze in an increasing visual world. Wearing his yellow bracelet, you tell everyone that you support Lance Armstrong and his campaign against cancer.

Yellow identifies you as a Livestronger, but bracelets now come in a rainbow of colors that tell people you support our troops or tsunami relief. The coalition of nongovernment organizations hoping to "make poverty history" chose white for their bracelet, which is interesting because their program is addressed mostly to black Africa. Sports teams have taken up the craze. Wear a blue bracelet if you are a Cubs fan—stamped with "Believe." Baseball teams sold more than 9 million color bracelets last year.

During the summer of 2005 the silicon rubber "message bracelet" hit the streets in a big way. It is all well and good that you can signify your team or cause by wearing a kind of solidarity bracelet. It is all color-coded as the visual increasingly takes over from the written word.

COLOR CONTROL

One of the hottest things in information management is data visualization. Companies deluged with information are using visualization systems to see data in pictorial form, with color becoming an increasingly important contributor. On July 7, 2005, the call traffic routes at T-Mobile in London started to get unusually busy. Alarms on operators' screens started to flash red. It wasn't long before the U.K. mobile telephone service management center learned that the huge surge in mobile phone usage grew out of the London bombings that morning. T-Mobile handled 86 million calls on its U.K. network that day and an estimated 500,000 text messages every 15 minutes. Through a visual representation system of what was going on, T-Mobile had no major faults in its servicing of calls.

In dealing with increasingly mountainous data, T-Mobile and thousands of other companies are using data visualization software that converts numerical data from spreadsheets and tables into pictorial, graphic, or animated form. This is part of a visual world, and it is the most important things that are visualized. "It is a constant battle to decide what is critical, what we really need to visualize," says Martin Pigott, T-Mobile's head of service operations. "Having too much information is far worse than having too little."

The many examples in this chapter describe how the eight forces are pushing into a visual world. Visual dominance will have consequences—

some direct, like the accelerated march of art into the consumer world; some unintended, like the decline in reading. Wherever you play in the marketplace, Mindset #11 becomes more and more important: Don't forget the ecology of technology. Innovation will fail when you don't listen to the heartbeat of our time. Pick up the rhythm or, in some cases, send enough adrenaline into the market to change that rhythm.

MINDSET #1 "While many things change, most things remain constant": It is not an either/or world. Both word and image will remain. But in many cases the written word will be replaced by visual representation and literary narrative displaced by illustration. Within the changing communication mix of word and visual, the visual will dominate. The challenge is to ascertain the optimum mix of word and visual in each field of endeavor.

MINDSET #2 "The future is embedded in the present": When you see changes occurring in the present, look for other signs pointing in the same direction. The eight main forces discussed in this chapter taken together are an example of how single pieces, observed in the present, can form a new picture of the future. The number of forces and their evidence vary, but never build your conclusion on a single signal.

MINDSET #11 "Don't forget the ecology of technology": The changes that result from the technologically driven onslaught of the visual can have profound impacts on the environment in which you operate and the environment of the marketplace that you address. Ask yourself what in those environments will be enhanced, what will be diminished, and what will be replaced because of the changes.

Chapter 2

ECONOMICS

From Nation-States to Economic Domains

The economic borderlines of our world will not be drawn between countries, but around Economic Domains. Along the twin paths of globalization and decentralization, the economic pieces of the future are being assembled in a new way. Not what is produced by a country or in a country will be of importance, but the production within global Economic Domains, measured as Gross Domain Products. The global market demands a global sharing of talent. The consequence is Mass Customization of Talent and education as the number one economic priority for all countries.

We need to reset and render more precisely our understanding of economic interrelations. In the twin paths of globalization and decentralization, what we have been witnessing is not the globalization of countries but the globalization of economic activities. And yet we perpetuate the fiction that the global economy is composed of the economic activities (GDPs) solely within each of 243 countries that are then totaled to tell us the size of the whole.

When we speak of the GDP of a country—the Gross Domestic Product, it is the sum of a national total of goods and services produced *within* a country's borders regardless of the nationality of the contributing companies. Thus the new $1 billion Hyundai plant opened recently in Mississippi will produce automobiles that will be counted as part of the United States' GDP, not as part of South Korea's GDP. The GDP of each country is treated as a closed system—which virtually no country is—and it is then announced that the country's "economy" has grown 2.1 percent or declined 1.6 percent. The very precision of the figures is something of a joke.

Those figures as the score of the game of economic developments will no longer deliver a meaningful result for a person or a company in any country.

The true *economic* borders and economic figures of the global economy should be drawn around global economic activities—such as the production and distribution of automobiles, which involve virtually every country in the world—rather than around countries, each of which is involved in only pieces of global economic activities.

I call these proposed areas of economic-specific activities *economic domains*: the automobile domain, the pharmaceutical domain, the financial services domain, the tourist domain, and so forth.

Although Germany as a country is stumbling along at little or no growth and has an unemployment rate of around 10 percent, German companies operating globally within their domains—companies like BMW, Siemens, and Adidas—are doing very well. The exports of German companies increased more than 7 percent in 2005.

Measuring the world's economy by following economic activities without regard to country borders does not mean that countries will not be thought of as countries. On the contrary, in the process of becoming more and more economically *interdependent*, countries will enhance their identities by becoming more culturally nationalistic, a subject I write about later in this chapter. But in collecting the pieces of the new economic picture of the world, the focus has to be on what is happening in the domains rather than what is happening in the countries. I am not talking about the theme of the political left in the 1970s and the 1980s that the "multinational corporations" were taking over the world. I am talking about economic activities of all kinds taking over the world, replacing geopolitical activities.

COUNTRIES DON'T CREATE ECONOMIES

In recent years as we have moved through globalization, we have begun to speak of countries in economic terms: the United States as the world's leading economic power; Japan, the fading economic star; Germany, Europe's biggest economy, in a period of no growth. We don't wonder what will happen to Germany but ask what will happen to Germany's economy. We speak of the great dynamics of China's economy—and speculate about when it will "overtake" the United States. When will Brazil become the great economic power it has always had the promise to be? And when will Russia finally have a free-market economy?

Of course all of this is either shorthand or nonsense, because countries don't create economies. It is entrepreneurs and companies that create and revitalize economies. The role of governments should be to create a nourishing environment for entrepreneurs and companies to flourish, not to get in the way of economic development. (Most governments think their role is to create rules and regulations. Microsoft or Google never could have developed in France. And now in the European

Union, rules and red tape are being transferred and magnified by Brussels.) The repositioning of the role of governments is embedded in the ecology of a global market.

EXPLOITING OPPORTUNITIES IN ECONOMIC DOMAINS

When an American or a German asks, "How is my country's economy doing?" the real worry is, "How is tourism doing?" or "How is the car industry doing?"

For the Fiat worker in Lyon, or the Ford worker in Detroit, it is how Fiat or Ford is doing, regardless of how France or the United States is doing. And now Fiat is a global enterprise, as is Ford. In turn, Fiat and Ford together are part of a huge global automobile sector, or domain; all the companies—car framers, engine makers, battery makers, glassmakers, distributors—are interconnected, constituting a gigantic automobile domain (without borders) that makes the world's cars. As a global economic phenomenon, the automobile domain is much more transparent than any country. It is much easier to know how an economic domain is doing than how Germany or France is doing or, even harder, how the global economy works.

If we focus on the score of the game, we have to know what game it is whose score we are seeking. Today's GDPs used to be GNPs—gross national products. The old GNP index was the sum of all things produced in, say, Germany, *plus* all things produced in other countries by German companies. (GNP = a national total output of goods and services produced *by* a country in one year.)

As the globalization and integration of economic activities increased, this became just too complicated to keep track, so about 15 years ago the GNP was given up by the commerce departments of most countries, and in its place we got the GDP. (GDP = monetary value of all goods and services produced *in* a country in one year.) GDPs count only those products and services created *within* a country's borders, no matter the nationality of the company. So integrated are the activities of American companies with those in the rest of the world that it can already be said that there is no such thing as a U.S. economy.

The only thing that will eventually matter will be the GWP—Gross *World* Product—of a single world economy, and the only way to understand and measure the global economy will be by examining selective gross *domain* products—a new GDP, but for domains.

Germany as a geographic and political entity does not create the German economy. Germany's economy is created by German companies and German entrepreneurs, many of them part of a larger economic domain, a global economic sector. Europe's economy is created not by the geopolitical European Union but by companies in Europe, some linked with other companies all over the world. Thinking and speaking of a German economy is a *conceptual fallacy*. There is now no such thing as a German economy—unless you arbitrarily count just what is physically made within the boundaries of the country and declare that that is the German economy. The German economy is so integrated with the rest of the world that there is no practical way to sort out what is "German" and what is not.

There are millions of small companies that in turn are part of larger economic domains, just as cities and towns are part of countries. Indeed, the fastest growing business sector in the global economy is individual entrepreneurs on the Internet.

Entrepreneurs in all continents are looking to make the most of business opportunities globally, instead of being stuck with local problems. A hundred years ago, the emperor of Austria and his royal household could enjoy his favorite chocolate specialty, produced in a fine bakery in the Alps, only when he was in his summer palace in Bad Ischl. Today via the Internet, the same chocolate is offered and distributed to chocolate gourmets all over the globe. Vietnamese artists work in their homes to produce paintings that are sold via the Internet to the world. A knife company in Vienna lists its products on the Internet, using DHL or FedEx as its distribution system. An entrepreneur in California put up his Web site 1daybanner.com and guarantees customers worldwide a banner of any size within one day after the order is confirmed, a great boost for demonstrators in places short on banner makers. eBay, one of the biggest global companies, created a huge global network of buyers and sellers and is now doing more than $10 billion a year. In my own case, with contracts to publish my books and make speeches, I have been operating for years in more than 50 countries with only a couple of associates, my global domain. Mine is a simple and now fairly old model, but with greater enabling technology, an increasingly important domain model. We are all a part of larger economic-specific domains—publishing, food, auction, and so forth.

In the beginning of the 1980s in Québec City in Canada, a city with no circus tradition, a group of ambitious young stilt walkers, fire-eaters, and jugglers created the High Heels Club. In the process of their developing their program, they decided to create a festival to exchange ideas and talent with other groups. Four years later, in 1984, the Cirque du Soleil was born. Their aim was to get past traditional rules and to defy the reality of limits. It reinvented the old institutional circus by showing a dream world, a dream that could be shared around the globe. It has made a triumphal march across the globe and today is an astonishing $2 billion part of the entertainment domain.

In tiny Finland, entrepreneurs created today's Nokia, whose focus on the global domain of mobile phones and awareness of the importance of great design conquered the world.

Some participants are seeding the market in quite inventive ways. Students in the seaside town of Arhus in Denmark volunteer to populate the world. Every day dozens of male students pop into Cryos International Sperm Bank Ltd., the world's largest sperm bank, to make anonymous deposits. Cryos ships the frozen sperm to customers in more than 40 countries. One student whose sperm has been sent to several countries has sired more than 100 children so far. One can only speculate where all this is heading, but it is a good example of creating your own global domain. (I got my Danish genes the old-fashioned way; my mother is Danish.)

Globally there is a vast collection of uncountable domains, not struggling to know how the global economy works or how their home country's economy is doing but doing what works for them within their economic domain.

THE DOMAIN WORLD IS LIKE SPORTS

The easiest way to reset your thinking is to start with a sports domain. Imagine that you want to become a football player. What is of importance? The best training and education to become a good player, no matter in what position you play. To gain as much information as possible about what is going on in the football world: which are the best teams, players, coaches? The best teams will choose the best players, and the best players will be able to choose their team. Your success and position within the football domain will depend on talent, ambition, flexibility, and commitment to your goal.

You handle networking within your football domain like the global telephone system handles phone calls. There are 2 billion telephones in use in the world today, and you or I can call any one of them, and each of them can choose to call you. But we don't. Within the larger system, we each create our own domain of telephone numbers we call or are likely to call, otherwise ignoring all the rest of the larger system.

Not knowing about other sports does not diminish your participation in your sport. Not knowing about other parts of the global economy does not diminish your participation in your global domain. Knowing what is relevant for your sport or domain and the fields linked with it will give you the competitive edge, but you don't have to know about any of the other global domains.

Countless self-sufficient domains make up the self-organized global economy. Globalization is a bottom-up phenomenon with all actions initiated by millions of individuals, the sum total of which is "globalization." No one is in charge, and no one can anticipate what the sum of all the individual initiatives will be before the result is manifest. A global economy can only be the result of "spontaneous order."

We have to move from thinking about countries that house enterprises to thinking about the enterprises themselves becoming the basic units of the global economy.

All powerful domains have one thing in common: Each of the parts functions as if it were the center of the network. When you are on the Internet in your home somewhere in the world, sending and receiving messages with others all over the world, you experience being in the center, just as do all participants in global networks. That is very powerful.

IS WORLD GOVERNMENT NEXT?

Could you imagine voting for a president of the world? Many people ask me if globalization will lead to a world government. God, I hope not. Why would we add a world government at a time when we have been subtracting power from the hands of national governments through privatization and global communications? The NGO (Non Government Organization) domain, a recent invention, has been organizing globally around economics and beyond, and has contributed greatly to the global decentralization of political power. In the 1980s, there were very few NGOs. Now there are millions of them all over the world, increasing the density on the global stage, and free of constraints of national borders. A widespread business complaint at World Economic Forum meetings in Davos, Switzerland, is that the NGOs had taken over the agenda (global warming, poverty, corporate social responsibility, and so on.)

The shift to economic domain thinking is all around us. If you are in the financial sector, you clearly understand that you are operating in a financial domain without regard to sovereignty considerations. Look at the currency traders. They pay absolutely no attention to country borders. The world is their domain.

Perhaps an economic domains report could be started as a companion index to the GDP reports. In time they would supplant GDP reports because GDP reports don't serve us in knowing what is really going on in the global economy.

I don't think it is too grand to say that thinking in global economic domain terms requires totally new economic models and totally new political perspectives.

Once you have reset your thinking and begun to collect pieces of the new picture puzzle, you will see the signs pointing in the direction of global domains in many fields. One of the forces supporting economic domains is decentralization.

DECENTRALIZATION IS THE KEY

The force that stands behind the development of economic domains is decentralization. The paradox is that it is decentralization that will make globalization work.

For several decades, we have witnessed globalization gain and centralization decline. We have witnessed this decline of centralization as an organizing mechanism in many parts of the world, most dramatically with the fall of the Soviet Union in 1991.

The worldwide privatization that began in the 1980s also represented a retreat from centralization, from nationalized industries, from government to the private sector. Widespread reduction of central regulations followed. It is clear to me that to be successful, to continue growth and to be sustainable, *globalization is wholly dependent on the degree to which the world continues to decentralize.*

It is a shift to a new appropriate scale, right down to the individual, an interplay among millions of individuals and companies, economic domains each acting in their own interest. Growth and stability in the world are wholly dependent on the degree to which the world decentralizes.

Note how this decentralization of nations is now symbolized by the actions of more than 20 countries—so far—to promote the appearance of individual citizens on their postage stamps. In Canada, Australia, Singapore, Austria, and the United States, among others, you can buy authentic postage stamps with your image on the stamps. Not the image of the king or the queen or the president or prime minister but you and your next-door neighbor.

I want to borrow something from the scientific world. In science, the particle physicists, who study the smallest structures in nature, have linked up with the cosmologists, who study the universe as a whole. A similar phenomenon occurs in the worlds of politics and economics where the particles and the cosmos are bound together—decentralization and globalization. And the interplay of the particles and the cosmos is a model for the study of the global economy.

BUSINESS DECENTRALIZATION

As we all know, within business, information technology is leading to greater decentralization, more flexibility and creativity. In a sense, many companies are reconstituting themselves as networks of entrepreneurs. But we are still in the early stages of human freedom in business that might in time be as important for business as the change to democracies was for government.

In the years ahead, the more manageable the pieces, the better the world will work. Computers allow us to organize and keep track of complexity, the complexity of having many small units—for companies and for the world. Computer systems allow us to decentralize control, to enhance the contribution of individuals all along the information chain, to make more of the talent we have. For a long time, Microsoft has been essentially a network of software programmers who communicate directly with Bill Gates.

On eBay there are more than 430,000 people who make their primary living as eBay sellers. (If they were a company, it would be one of the five biggest companies in the United States.) These almost half million people are in a real sense self-employed proprietors of their own retail stores, with a global market for their products—together doing more than $10 billion worth of business.

There is also a decentralization of titles going on in business: many presidents, many CEOs. GE now has 8 or 9 CEOs. Bertelsmann, the German media giant, has 10 chief executives in the United States alone. Besides providing the advantages of decentralization, it is a great employee incentive strategy: Because there is not just one CEO but as many CEOs as there are divisions, employees can strive to work their way up in their division. Bill Weldon, chairman and CEO of fabulously successful Johnson & Johnson, says about the size of his company, which is moving toward $50 billion a year in sales:

We don't think of a $40 billion–plus corporation. We think of 200 smaller businesses and each one of them is focused on the market they compete in. I describe it as if you want to push 200 grains of sand up a hill. It is a lot easier than trying to push a 200-pound boulder up. If that boulder backslides, you're in trouble. If you centralized and cut out infrastructure, couldn't you be more cost-efficient? Sure you could. But what about what you give up to take some costs out? Do you give up that ability to beat the law of big numbers?

The bigger an entity of any kind, the more it must be made up of smaller and smaller, more efficient parts to be successful. Big companies are decentralizing not because it is the fashionable thing to do but because they have to become more efficient and responsive to the new competition generated by the admission of almost everyone to play in the new global game. That's why GE now has 8 CEOs. That's why Johnson & Johnson now thinks and acts as 200 companies (200 CEOs). It is embedded in today's world that in the future, business success will be determined by innovation and swiftness to market. By these lights, small, agile companies, or companies that have reconstituted themselves into confederations of entrepreneurs will beat big, bureaucratic companies 10 out of 10 times.

THE WINNER IS SMART, SMALL, FLEXIBLE

The decentralization of companies mirrors the increased success of small, swift-acting entrepreneurial companies.

A small Austrian team saw an opportunity in the growing gap between advances in surgery and the lack of technological support to become a global supplier of high-end microscopes for surgery. A new generation of surgeons had changed surgery and invented refined and complex surgical techniques, allowing them to operate flexibly in micro-dimensions. What was missing was adequate operating microscopes.

The company Life Optics invented the Varioscope, a small micro-scope worn like glasses that offers autofocus and zooms together with integrated lighting and charge-coupled device video camera, all of it in the size of a small loupe. It replaces the big, bulky, inflexible, and—most important of all—nonadjustable traditional surgical microscopes. It en-ables surgeons to more easily use complex and refined surgical tech-niques in, for example, prostate operations, where they can avoid the common consequences of impotence and incontinence, a great boon for the patients.

None of the big players in optical electronics can get around the Life Optics patents. Life Optics has succeeded in creating a new worldwide domain and is asserting its place in it. The much better results of surgery support the more widespread use of the Varioscope, now in more than 60 countries. Once the benefit of the instrument is revealed, the change to the new equipment is only a question of time.

THE FALL OF THE LAST MONOPOLY—THE PRIVATIZATION
OF NATIONAL CURRENCIES

Today, no domain is as firmly in one hand as the domain of national cur-rency, untouched by all privatization programs of the last decades.

Beginning in the 1980s, Margaret Thatcher initiated the privatiza-tion-of-businesses program that has been copied around the world. The rush and extent of the global privatization of businesses—including util-ity monopolies, telecommunications, postal service, railroads, railroad monopolies, and banking monopolies—was one of the most extraordi-nary trends we have ever witnessed. Now that government monopolies have been privatized and made competitive worldwide, it is time to abolish central banks. What is now needed for the future development of the global economy is competitive currencies, currencies that are directly competitive with national currencies, just as we needed competitive phone companies to directly compete with national phone companies.

In the United States between 1837 and 1866, there was what is now referred to as the Free Banking Era. Paper money was issued by, and backed by, states, companies, private banks, and even individuals. There was a true competition in the market.

There is a wave of new currencies flooding the world. There are now more than 4,000 communities around the world that have created their own currency for social purposes, from the 400 private currency systems in Japan to pay for elderly care to Ithaca Hours, in Ithaca, New York, a paper currency system, where time is credited in an exchange of social goods and services.

These "complimentary currencies" include frequent-flyer miles, the currency issued by airlines, which now can be used to pay for telephone calls, taxis, restaurants, hotels, and merchandise—even if you have never flown in an airplane.

And we also have "alternative currencies," such as credit cards and other electronic exchanges that are growing rapidly. Frequent-flyer miles coupons could also be called alternative currencies because you can use them in place of money. (The *Economist* has calculated that the total stock of unredeemed frequent-flyer miles is now worth more than all the dollar bills in circulation around the world.)

During the 2003 Venice biennial toward the end of our stay, my wife Doris and I visited the Taiwan pavilion that was housed in the Palazzo delle Prigioni, just off Piazzetta San Marco. On entering, we saw that much of the floor was covered with beautiful, plump, stark-white garlic bulbs. As it turned out, they were advertisements for a new currency.

We were invited into the G-Mart: G for *garlic*, the notes issued against farm-grown organic garlic, "a commodity that is desired and wholly revered in the year 2030, when garlic will become the world's currency of exchange." Chinese artist Shu Li Cheang said that his installation "is a foil to our present economic system, raising questions of the value and meaning of money." I still have several hundred G-Mart 10,000-unit notes, all redeemable against organic garlic.

The world is creating a virtual kaleidoscope of currencies, complimentary and alternative. But complimentary and alternative currencies are wholly subsidiary to national currencies issued by governments. These national currency monopolies remain unchallenged, but not forever.

The last step in decentralization will be the privatization of the Mother of all Monopolies, national currencies. The mindset breakthrough is to understand that money is a commodity like cars, refrigerators, potatoes, and gold.

Government currency monopolies have failed. Politically, the failure is built in. Currency monopolies have the defects of all monopolies: One must use their product even if it is unsatisfactory, and above all, it prevents the discovery of better methods of satisfying a need to which a monopolist has no incentive to respond.

A transparent monopoly example was the East German Trabant, a famously primitive monopoly car, brought to life by a shortage of steel and fuel and affectionately called the Trabi. It was "the epitome of socialism: bad, not capable of doing too many things, not too efficient but still manages to work somehow." It was the only car East Germans could buy. The East German government determined the price, and there were almost no improvements made to the car. Nevertheless, it was the dream of many in that Socialist and Communist country, where nothing could be taken for granted, where as late as in the early 1980s crowds flocked to stores when news came that bananas or oranges had arrived. Indeed, the East German economy itself looked like a Trabi.

People generally are not aware that the currency monopolies are not much better than a Trabi monopoly. It has long been a common belief among economists since the classic thinkers of the eighteenth century that one of the most important functions of government was to create a monetary mechanism and to issue money. (By the way, it was not among those duties that Adam Smith said fell to the state.) Now this common belief is coming under challenge.

If we take time to have a closer look, we easily discover that governments will continue to fail to supply good money. And the purchasing power of all government monopoly currencies will continue to go down—at the expense of all citizens:

- From 1971 to 1991, through inflation, the German mark lost more than 52 percent of its value, and Germany had the *best record* of that period.
- By 1991, the U.S. dollar had lost more than 70 percent of its 1971 value.
- The British pound lost more than 84 percent of its 1971 value.
- Within the United States, the value of the currency between 1995 and 2005 dropped 33 percent. In the 10-year period from 1995 to 2005, it dropped almost 20 percent.
- Take any period of any currency monopoly, and the result is the same.

And we all pay the price. Everybody pays for the inefficiencies of currency monopolies.

Paul Volcker, the former chairman of the Federal Reserve, said in a 1990 speech that central banks were "not on the cutting edge of market economy," that "central banking is almost entirely a phenomenon of the 20th century," and that when established, "central banks were looked upon and created as a means of financing the government."

For a long time, it's been said of governments that "they will just print money." Now, they create money digitally as well. Technology is the great enabler for new currencies and currency systems.

We cannot be sure where it will all lead. But a new freedom of choice seems to be the banner. The choices will go beyond what we can now imagine, if there is the freedom to find them and explore them.

The good news is that the process has begun to end government currency monopolies. And any dose of strong inflation will stimulate more interest as governments take even more of our currency value. So think outside the monopoly currency box. The future of money is complimentary, alternative, and competitive.

Alan Greenspan, whose term as chairman of the Fed ended in 2006, said, "We will see a return to private currencies in the 21st century." That's a big time frame, but I'll take it.

THE GLOBAL SHARING OF TALENT—SPORTS AGAIN IS LEADING THE WAY

Hand in hand with decentralization and global economic domains is the global sharing of talent. Outsourcing, about which we have heard so much recently, is the new handmaiden of global decentralization. It puts the recruitment of individual talent on a global level, decentralizing right down to the individual.

A few years ago when Real Madrid was arguably the best soccer team on the planet, it had only two Spanish players. European soccer is notorious for having teams with more foreign players than domestic. Today there are more than 100 Brazilians playing on European teams. Soccer teams are a model for outsourcing highly skilled talent.

The United States is pretty good at it, too. On opening day of the 2006 baseball season, more than 30 percent of all major-league players were foreign born. In the minor leagues, an astonishing 50 percent of the players are foreign born. (This is outsourcing big time.) And basketball is spreading all over the world. NBA games are broadcast in 205 countries and 43 languages. The country with the most foreign-born players in America's NBA is Serbia, with 10. In the 2005 season, 15 percent of all starters were foreign born, and six of these were in the all-star game. I think it could be said that today half of all the best players in the NBA are foreign born.

The San Antonio Spurs, who won the 2005 NBA chamonship, are also the most international team in the NBA. They, in fact, have more foreign-born players than American players. Seven of the 12 players allowed per team are foreigners from Argentina (Manu Ginobili and Fabricio Oberto), France (Tony Parker), Slovenia (Rasho Nesterovic and Beno Udrih), the Virgin Islands (Tim Duncan), and New Zealand (Sean Marks). They also have rights to probably the best big man in Europe, 6′9″ Luis Scola from Argentina, who is now playing for a Spanish-league team, Tau Cerámica. On June 26, 2006, an Italian 6′11″, 225-pound forward, Andrea Bargnani, who has been playing for the Benetton team in Europe, became the first foreign player to be selected number one in the NBA draft.

Embedded in the sports model is a preview of what will happen in economic domains. Supply and demand determines the market. The teams choose their players, and the best players choose the team they want to play for. As I wrote earlier, positioning within a domain is everyone's own responsibility.

Economic problems in certain geographical locations can be balanced out by exploiting the opportunities of the global game. Change should be embraced when benefits can be achieved.

THE MASS CUSTOMIZATION OF TALENT

In the business world today we can draw on talent from many levels and many diversified skills from wherever we find it in the world, just as has been done for a long time in sports. We hire someone with the best skills from another company to strengthen our company, and we can do it all electronically and globally. Building on Stan Davis's concept of mass customizing, I call this era we are moving into the mass customization of talent.

We are taking Adam Smith from the division of labor to the division of talent, from mass customization of shirts and jeans to mass customization of talent. Individual talents fitted to needs—globally. The sports team is the model, elevating outsourced cheaper manufacturing labor to outsourced expensive talent. Now, instead of football players, think engineers, physicians, designers, scientists, administrators. The world is regrouping at a higher level.

Today, the strategy to prevent getting bigger and more bureaucratic is more domestic and offshore outsourcing—including such previously sacrosanct functions as human resources management. (British Petroleum's human resources functions are now entirely handled by a firm in California.) Four months ago, giant chemical company DuPont outsourced its human recourses—including workforce planning, labor relations, and performance management—to a firm called Convergys Corporation of Cincinnati, Ohio. The 13-year contract covers 60,000 employees and 100,000 pensioners in 70 countries and is worth $1.1 billion.

This development offers amazing opportunities for medium and small companies to serve larger companies, even though the mantra of CEOs and human resources directors has for years been "Our human resources are our most valuable asset" (but we are now willing to pay strangers to manage them).

Just as the Houston Rockets, an American team, outsourced Yao Ming from China to play basketball, Sony in Japan outsourced American Howard Springer to be its CEO. The former CBS executive doesn't speak Japanese, and he doesn't even have an office in Japan. Springer's response to criticism about not moving to Japan or learning to speak Japanese was "I don't live in Hollywood, but I am responsible for moviemaking. I don't sing songs, and I am responsible for a music company."

OUTSOURCING CAN BE ACROSS THE STREET OR ON THE OTHER SIDE OF THE WORLD

In Missouri, generally not considered to be an epicenter of innovation, McDonald's has made a 900-mile step to improve its service. When a customer stops at the order window and places his wishes for an opulent burger meal, he would never think that his order is traveling 900 miles to Colorado to an order taker who takes an electronic picture of him, displays the customer's order on a screen (to make sure it is right), then forwards the order and the photograph back to the Missouri kitchen, before the customer has even started to move toward the pickup window.

Are workers in Missouri losing jobs to people in Colorado? Yes, of course they are, just as workers in Colorado are gaining jobs. And customers in Missouri are getting better and faster service.

It is not only McDonald's hamburger orders that go on a service journey. Radiologists in India are analyzing the X-rays—sent over the Internet—of American patients. It reduces health care costs; it raises American health care productivity. Who could object to that? Yes, I know, American radiologists!

If an X-ray can be analyzed cheaper in India, nothing will stop doctors anywhere from using Indian radiologists. If an automobile can be made more cheaply in Korea, it should be and will be. If a telephone inquiry can be processed more cheaply in Ireland, it should be. It raises incomes on both sides because resources are advantageously redeployed. Trade is a positive-sum game, as Adam Smith was the first to understand.

Of course, massive offshore outsourcing of manufacturing jobs has been going on for decades. What is beginning to happen now is offshoring of talent at much higher levels to obtain skills and knowledge of much greater diversity. The transplanting of work done by highly skilled professionals is a change in the terms of trade in global competition. The global trading system is regrouping at a higher level. We have talked about this for a long time, but now companies can and will seek to bring together the best skills from around the globe. Not, let's get the best team of engineers in all of the United States for our project, but, let's get the best team of engineers in the world. As of now, the top offshore sourcing destinations are, in order, India, China, Malaysia, and the Czech Republic. At the higher-skilled end, we should add Singapore, Canada, and New Zealand.

The answer for everybody regarding this extraordinary development is better education and training. That's been the answer for a long time, but now we all really have to get serious about it. The United States has a big lead in the increasingly competitive global economy, but it could lose it if it doesn't become serious about its basic school system. For years the quality of its graduates has been going down, and as in the U.S. steel industry, the subsidies keep going up as the product declines. Elementary and secondary education in the United States is a declining industry, its products performing less and less well. It is a monopoly, propped up by subsidies, protected by strong unions, and with poor management with no real accountability. The situation is sufficiently dire that I am left only with the hope that it will be privatized and that competition will work its corrective magic. The reason the U.S. higher education system is the best in the world is because it is a competitive system.

For years I have urged countries that their number-one economic priority must be education and training. If you want to be a player in the global economy, your number-one economic priority must be education, as an individual or a country. The primary role of education has been one of the constancies in the past, and it will continue to be so in the future.

My wife Doris and I were in Vietnam recently, where we witnessed the country's big push to turn itself into a major global outsourcing power. In Vietnam, mathematics has been a strong part of the education curriculum (the one thing the Communists were good at was education), and it is now pushing a strong computer curriculum. And salaries are very low. World'Vest Base, a Chicago company, has 38 employees in Hanoi who surf the Web for the action on obscure stock markets, information that they pass on to clients in various parts of the world who specialize in emerging markets. They start their young employees who have degrees in finance or accounting at a salary of $100 a month.

In Saigon, Atlas Industries has more than 100 employees who turn architectural drawings into blueprints for British construction companies. Vietnam sees itself tooling up its outsourcing to compete directly with India in the near future.

THE END OF CHEAP LABOR

We know that chasing after cheap labor outside a country's border is coming to an end, for two reasons:

1. Salaries are climbing dramatically in emerging economies, and wages, in any case, reflect productivity. (Mexico's productivity is

one fifth that of the United States.) And it is axiomatic that the more world trade there is, the more wages level out.

2. The labor *content* in manufactured goods is declining dramatically: In 1970, it was around 25 percent; today, it's about 4 percent.

DOES GLOBALIZATION MEAN AMERICANIZATION?

The twin directions of globalization and decentralization are leading us further away from a world of nation-states to a world of economic domains, not as a new way of organizing the world but as a new way of understanding the world. Two prejudices stay: Asia is taking away jobs, and America is undermining national culture. That is not the future I see embedded in the present.

As often as I listen to the worries about China eating the jobs of the West, I hear the concern about the influence of the American way of life in the East. The question is: "Does globalization mean Americanization?" My short answer is no. In measuring globalization, we can count telephone calls, currency flows, trade sums, and so on, but the spread of culture and ideas cannot be so easily measured. Embedded in the present is the unrecognized paradox that culturally, America itself is changing more dramatically than America is changing the world. It is the world that is changing the world. Immigration is reshaping America more profoundly than America's influence around the world.

The so-called Americanization of the planet that people talk about and rail against has to do only with the most psychologically superficial of decisions: what we eat, the movies we see, the jeans we wear. And think about this: In the United States, there are more Chinese restaurants than there are McDonald's.

America is already by far the most ethnically diverse country in the world. Forty years from now, 100 million of the 400 million people who will be living in the United States will be Latino. By then, Latinos plus Asians will make up one third of America's population. Today, Islam is the second biggest religion in the United States after Christianity. The second most spoken foreign language in America today is Chinese. These shifts amount to profound change in who Americans are and what America is.

For the most part, people all around the world are working to strengthen their cultural identities. The paradox is that the more universal we become, the more (consciously or unconsciously) concerned we are about our cultural identity. This is one of the constancies against all the prophecies of homogenization. All over the world, the things that are truly important to each of us are in our bones. Our mother tongue, our family, our community, our cultural heritage, and the strength with which we hold them differs from country to county and person to person. The role of nation-states will increasingly be to celebrate cultural identity and inheritance, and to attend to education.

Economics will continue to overwhelm politics in the running of the global economy, and the impact of the global economy on our lives is far greater than the rhetoric of politics. The future is embedded in today's economic decisions. That's why we need better indices to keep track of the economic globalization of the world. The economic relationships and activities of the world are integrating and self-organizing. The window through which to view and understand the economic integration of the world is economic domains.

Applying Mindsets

MINDSET #5 "See the future as a picture puzzle": Anticipating the future sometimes begins with a feeling. In recent years I began to notice that economic indices, based on Nation States, increasingly lost practical value. The economic weight was shifting from Nation States to what I now see as Economic Domains. To anticipate economic developments, the pieces to keep an eye on are domains.

MINDSET #10 "Don't add unless you subtract": We tend to draw a much too wide circle around what we think we have to know. Not everything that is going on in the global economy is relevant for us. The circle of interest should be drawn around our Economic Domain. Because domains operate independently, frame the field around your own Economic Domain and then don't add unless you subtract a less important consideration.

MINDSET #2 "The future is embedded in the present": Information about Economic Domains is much more available and reliable than information about the GDPs of countries. This greater transparency will help us create a truer picture of global economic activities. At the beginning, the Economic Domain Index could be a companion index to the outdated GDP index, eventually replacing it.

Chapter 3

CHINA

The Periphery Is the Center

Many people worry that China may break up. Well, China is breaking up, walking the twin path of globalization and decentralizing more than any other country in the world, a process essential to China's sustainability, giving more efficiency and power to its parts—cities, provinces, and regions. The periphery is the center.

When Deng Xiaoping visited the United States in 1979, he was taken on a tour of the Ford plant outside Atlanta, a factory that at the time made more cars in one month than all of China produced in a year. While in the U.S., Deng said, "We in China are faced with the task of transforming our backwardness and catching up promptly with the advanced countries of the world. We want to learn from you." He said he hoped to transform China into an industrial power by the year 2000, then 21 years away. In 1979 China manufactured 13,000 cars. In 2004, 25 years after Deng's visit to the United States, more than 5 million cars were produced in China by 120 carmakers.

Deng began his agricultural reforms in 1978, and he journeyed to the United States the next year to celebrate the normalizing of relations between the two countries and to announce that China was open for business. I remember attending a reception in Washington for him during that visit and being somewhat stunned, as many others were, by what a tiny man he was, a tiny man with very large plans.

My own experience with China started 40 years ago. It was during the time I was living in Thailand, helping to accelerate agricultural development. One of the managers of a company I was working with took me on a trip to Taiwan. Taiwan was just at the beginning of its industrialization, but its agricultural reform measures were impressive, and I was going there to learn how they did it. As I write this, I remember a meal we had in T'aichung in 1967. It was a special place my friend had chosen, but even Taiwan's most modest current restaurants would beat the quality of that food by far, not to mention the hygienic considerations. During the meal, as luncheon entertainment, a grim-looking man who could have just stepped out of a Genghis Khan movie puffed himself up in front of our table and began his fire-eater performance. Just next door, hundreds of almost naked workers were excavating for a building's foundation. The dirt was dug out by hand with shovels and carried away on the workmen's shoulders in baskets. Everything in Taiwan was in an almost incomprehensibly primitive state.

Today, the further I look back in my memories of China, the more I am conscious of what extraordinary changes have taken place. Again and again, in each of my three to four visits a year, I am surprised and impressed.

UNDER THE WHEEL OF HISTORY

In 1957 Mao Tse-tung said: "The socialist system will eventually replace the capitalist system; this is an objective law independent of man's will. However much reactionaries try to hold back the wheel of history, sooner or later revolution will take place and will inevitably triumph." That was his understanding of what was embedded in the present. At the time of my first visit to Taiwan, mainland China was still under his iron rule, a man who controlled almost a billion people for more than 25 years in a country of more than 5,592,000 million square miles, about the size of the United States. He overthrew an army of more than 4 million to get it, and killed many more to keep it.

Nixon started diplomatic normalization with China in 1972, and sometime after, I was able to make my first visit to the mainland. It was not easy, and every visit was under control of the authorities, who had very little tolerance for visitors who wanted to move about. My memory is without color: less than modest houses, gray laundry hanging on rods, and people in gray Mao suits filling the streets. I never stepped on a Chinese plane to go from one Chinese city to another. As did many others, I flew only from and to Chinese cities from British Hong Kong, where I based for any exploration of the mainland.

After Mao's demise in 1976 and the battle over the country's leadership, Deng Xiaoping finally became party chairman. He became the great change agent for China, with the aim of putting market economic policies in place, a grand sweep of economic reform. His famous aphorism guided the way: "The color of the cat shouldn't matter so long as the cat catches mice." And the socialist cat, disrespectful of Mao's 1957 prediction, wasn't catching any mice when he took over, so Deng traded in the cat.

During the 1980s Deng started with special economic zones in selected cities. Going from those primitive days to today's supermodern

Shanghai, which in the 1920s was one of the major cities of Asia, is the current incomprehensibility. Deng, who died in 1997, surely will go down in history as the emancipator of China from the shackles of Maoism.

CAPITALISM IN CHINESE CHARACTERS

There are no Communists in China anymore, and the evolution of a modern China will in time lead to political reforms. Trust the process. The private sector is today the most dynamic component of the Chinese economy, growing at around 20 percent a year, more than twice as fast as the economy as a whole. The emerging class of entrepreneurs is propelling China's modernization.

Since 1995 Carrefour, the French supermarket retailer, has built up China's second-largest chain of stores. At the beginning of 2006 it had 70 superstores in China and is adding 15 more a year by ignoring the requirement for central government approval of retail joint ventures. It has made its deals directly with local city governments.

Beginning with my first visit in 1967, I have been witness to all these changes China has undergone. My associations there have been mostly with the leadership and entrepreneurs; I have lectured and met with students in many of China's universities, and I have been a member of the faculty at Nanjing University. Over the years I have also met with many members of the State Council (which in the United States equates to the president's cabinet) when they were in power, including President Jiang Zemin.

I met with President Jiang in 1998, shortly after *Megatrends Asia* was published in Chinese. It was during a visit in Beijing when I was told that the president would be interested to meet me. I gladly agreed. At that time it was beginning to dawn on all of us that something new was really going on in China. I met with the president in the Zhongnanhai, the compound of the leaders of the central government. He told me that the room where we were meeting was where Mao had received his guests.

During this two-hour private meeting, I learned a lot more about China and the man himself. In the context of economic growth, he asked me what I thought China's biggest problem was. I hesitated to judge China's situation, but he really wanted to know. So I said, "I guess the big problem is how to make China's big state companies efficient at a rate that does not create too much unemployment at any given time." A delicate balance, we both agreed.

Jiang continued, "We have a saying around here: that we grasp the big ones and let the little ones go." My entrepreneurial heart immediately said, "I hope I'm one of the little ones!"

The most memorable part of our meeting for me was during the second hour when we had a very personal conversation, talking about growing up—we both had grown up on a farm—and about our schooling. He told me that he had been an automobile assembly-line worker in Shanghai during the Cultural Revolution. All this was in English, at a time when no one in the West knew he spoke English. At one point, to my great astonishment, President Jiang recited for me Abraham Lincoln's Gettysburg Address by heart. As many of you know, it ends: "... that this nation, under God, shall have new birth of freedom—and that government of the people, by the people, for the people, shall not perish from the earth."

Imagine his saying those words. I did not know whether the words for him were merely the words of a speech, or if he somehow related to them. I still wonder.

WHERE DOES CHINA STAND?

China has certainly become the "workshop of the world" and is widely feared as the dragon that devours our workplaces with a satisfied smack. This is nothing new under the sun. Looking back in history, during the Industrial Revolution the world's workshop was Great Britain, later the United States, and then during the last quarter of the twentieth century, Japan and Korea. Now it is China.

But China will be not only the world's workshop, making more products than anybody. We will also see China developing technologies and creating new ones. It dominates the production side of the global apparel industry, exporting $95 billion worth of apparel in 2004. About 25,000 clothing designers work in China, the best centered in Shanghai, where there is a strong expectation that in a few years they can compete with the fashion centers of Milan and New York. Beyond workshop of the world, China over time will become one of the design centers of the world, not only in manufacturing but in fashion as well. Today we are exporting yesterday's jobs to China; tomorrow, China will be exporting some of the next generation of products to us.

WHAT IS EMBEDDED IN TODAY'S CHINA?

Remember, when we want to anticipate the future, an important mind-set is "Focus on the score of the game." This is not easy when it comes to China's economy. The buzz and hype about China's growth has led to great exaggerations of how fast it is growing, and many are predicting when its economy will overtake that of the United States.

In anticipating where China stands now and where it will be 10 years from now, you will have to bring China's extraordinary growth into the global context. China's newly revised 2004 per capita GDP still ranks lower than one hundredth in the world. China's economic base is very low compared with that of the United States, and the United States will not stop growing. Even at present rates of growth, it will be at least 30 or 40 years before China will catch up to America's standard of living. You also may want to keep in mind that "Things that we expect to happen always happen more slowly" and sometimes not at all.

Columnist Ben Stein wrote, in a comment on the China-will-overtake-us alarm, about the Soviet Union in the 1950s:

It reminds me a lot of how the news media and the Central Intelligence Agency went berserk after the launching of Sputnik in 1957, and it was forecast that the Soviet Union would soon be the world's technological and economic hegemony. That talk was based on a number of faulty assumptions and a good deal of hysteria. Obviously it didn't happen.

At one time it was the Soviet Union that was supposed to rule the world, and now it's China. What is missing, among other things, is any caveat about straight-line extrapolation.

This idea of China taking over was supported when on December 20, 2005, China announced that its economy was quite a bit bigger than previously estimated. The new figures suggested it had probably passed France, Italy, and Britain to become the world's fourth-largest economy after the United States, Japan, and Germany.

"We now have a new snapshot of the Chinese economy," said Hong Liang, an economist at Goldman Sachs in Hong Kong. "This is not slightly bigger—it's a significantly bigger economy."

China revised its economic picture after a yearlong national economic census—the first such census, which involved more than 3 million auditors and supervisors—uncovered $280 billion in hidden economic output for the year 2004. (Well, it's a big country.) That brought the country's GDP up to $2 trillion. The U.S. GDP is more than $12 trillion. Most of the $280 billion apparently came from a much stronger (from 32 percent to 41 percent) service sector and the increase in smaller and midsize companies. (As former president Jiang said: "Let the little ones go.")

The upward revision of China's economic status points to less reliance on manufacturing and exports than was generally thought, good news for companies wanting to sell their products. China continues to grow extraordinarily at almost 10 percent a year, as it did for almost three decades. That means its economy is doubling every seven and a half years.

Hong Liang said that "the most significant implication" of the announcement was "does China have some structural illness or cancer, or is there an error with the X-ray?" He said that during "the last few years so many famous economists cited the very high investment to GDP ratio as a serious problem. Now it looks like the X-ray machine had a problem, not the patient."

Today's question of economic supremacy is not a question of the United States versus China. It is eBay versus Jack Ma, founder of Alibaba.com and of the TaoBao consumer site in Hangzhou, 60 miles south of Shanghai. eBay's chief Meg Whitman says China is a "must win" if eBay is to be global. So she has taken on Jack Ma.

Jack Ma's TaoBao site and eBay are fighting tooth and nail for the Chinese market. Ma says, "eBay may be a shark in the ocean, but I am a crocodile in the Yangtze River. If we fight in the ocean, I lose—but if we fight in the river, I win." Analysys International said Alibaba's TaoBao. com leads the consumer-to-consumer market in China in transactions, and it will likely surpass eBay in subscribers in 2006.

TaoBao now has more than 40 percent of China's online auction sales, compared with eBay's 53 percent. At worker rallies, TaoBao employees wave flags imprinted with their mascot—worker ants, the idea being that tiny but united, they can defeat an elephant. [Late in 2005 Whitman bought a stake in TaoBao—if you can't beat them, join them.]

Ma's background makes him well suited to connect the Western and Chinese worlds of business. He grew up at the beginning of China's economic flowering in the 1980s, in the historic city of Hangzhou, near Shanghai. He was eager to learn and to improve his English, so he worked as an unpaid tour guide to foreigners visiting his city. Ma graduated from Hangzhou Teacher's Institute with a degree in English in 1988. For several years he taught English and international trade at Hangzhou Institute of Electronics Engineering. A partner like Ma is an incredible advantage for any company when it faces one of the challenges to establishing a brand in China—the barrier of the Chinese language.

THE PERIPHERY IS THE CENTER

While the West experiences its own competition with China as almost overwhelming, the competition within China is even more severe.

Almost all the talk about China centers on its manufacturing advances, while the changes in the urban landscape of China are underestimated and hard to believe unless you are there to witness its great cities' renaissance. Old urban centers are being torn down and millions of people are relocating in great construction projects that are shaping the China of the twenty-first century.

China has 166 cities with populations of more than 1 million—compared with 12 in Japan, 9 in the U.S. and 1 in Britain. There are many cities in China that one has never heard of that have populations of 6, 7, or 8 million people. The country's rapid urbanization has lifted hundreds of millions of rural Chinese out of poverty.

In 2002 Fred Li, the chairman of a Taiwanese company on whose board I sit, took us to a meeting with the local administration of a city whose name we only knew because it is on the label of a well-known Chinese beer: Tsingtao. Behind that name is a city with 7 million people and an astonishing economic growth of 32 percent the previous year. And as Fred had told us before, Tsingtao is one of the most beautiful cities in China. We stayed in a brand-new five-star hotel. Next to it, a Disney-like amusement park had just opened its doors. Across the modern highway, right on the Yellow Sea, luxury beachside villas were almost finished, waiting for affluent locals to buy or rent.

Almost all of these cities have been turned into vast construction zones. Every city wants to develop into a world city. They are all building their own international airports. The Western press focuses on China's three great cities: Beijing, Shanghai, and Guangzhou (we in the West used to call it Canton, but it is now "Guan-joe"). But across China, scores of reborn cities are, as the Chinese say, "sprouting up like mushrooms after a spring rain."

One of the leaders in this urban renaissance is the old industrial city of Harbin in China's northern corridor. Harbin was in the news in 2005 when a chemical plant explosion 200 miles upriver contaminated the water supply of the city. But that didn't slow down one of the most ambitious urban projects in China. Harbin is now building a new city center for 9 million people called Songbei, after the Songhua River, on which it is located.

Songbei will be 285 square miles of office towers, residential high-rises, luxury villas, five-star hotels, shopping and entertainment complexes, trade zones, and industrial parks. It will be almost the size of New York City.

Just a two-hour train ride north of Shanghai, Nanjing is creating a 37-square-mile area as a new town. And so have other cities, including Ningbo, Yangzhou, and Nanchong. Zhengzhou is developing a master plan for a new 58-square-mile town. Changchun in the north is moving its business district to a new city center.

The model for these futuristic new city centers seems to be Pudong, the great new extravaganza on the east side of the Huangpu River in Shanghai where more than 200 skyscrapers have been built in just the last decade. Shanghai is now clearing 1,300 acres of riverfront land to relocate 50,000 residents and 270 factories, including the country's largest shipyard, in order to build a site for the World Expo in 2010.

There is ferocious competition among the second-tier cities (second-tier to Shanghai, Beijing, and Guangzhou). Globalizing as much as decentralizing, they are escaping the control of the central government in Beijing. As the Chinese say, "The mightiest dragon cannot crush the local snake." In China today, Beijing pretends to rule, and the provinces pretend to be ruled. The old saying that in China "the periphery is the center" is more true than ever.

All this rush to modernity leads many observers to get too far ahead of the parade, predicting that China is going to become the world's leading economic power next Tuesday morning. But extrapolating too far ahead of the parade in your expectations will blindside you just as much as staying too far behind. Hold on to that Mindset! To make a judgment on where China stands and where it is going, you have to evaluate the different pieces of the Chinese picture puzzle against one another.

THE NAME OF THE ROSE DOES COUNT IN CHINA

The Chinese spoken language distinguishes a high level of internal diversity. Many variants of it are different enough to be mutually incomprehensible. Depending on the classification system used, there are between 6 and 12 main regional groups. The most populous are Mandarin, Wu, and Cantonese, in that order.

Chinese characters are morphemes independent of phonetic change. Nevertheless, the orthographies of Chinese dialects are not completely identical. The vocabularies of different dialects have diverged. In addition, while literary vocabulary is mostly shared among all dialects, colloquial vocabularies are often different. Colloquially written Chinese usually involves "dialectal characters" that may not be understood in other dialects or characters that are considered archaic in standard written Chinese, a translation nightmare for the West. *Lost in Translation* turns from a memorable movie title into the unintended humor of unrecognizable brand slogans: American Express's "Don't leave home without it" turned into "Stay home with it"; Burger King's "Have it your way" into "Reject the socialist ideal"; Rice Krispies's "Snap! Crackle! Pop!" into "Break, crush, and explode"; Sprite's "Obey your thirst" into "You are commanded to drink."

Despite my believing in a universal design language, moving a brand to China is quite a challenge. And I say that even though I am very positive in my thinking. So much so that several years ago at an event in California, the person who introduced me said: "John is so optimistic that he thinks of the San Andreas Fault as the San Andreas Opportunity." Within all opportunities China offers—its language does hide some San Andreas Faults.

And it is easy to plunge into the Language Fault. The Web site of the American Chamber of Commerce in China carried a report from Baker & McKenzie on such a failure of a company, no smaller than Coca-Cola.

The story of how the Coca-Cola Company chose its Chinese mark has often been told. However, it deserves retelling here as the best example of how the fusion of both transliteration and conceptual methods should work. According to Mr. H. F. Allman, Coca-Cola's former China legal counsel, the company found that ahead of Coke's major marketing effort in China, many shopkeepers had made signs adopting any old group of characters that sounded remotely like

Coca-Cola without giving a thought to the meaning. One improvised sign translated it as "female horse fastened with wax"; another exhorted customers to "bite the wax tadpole." ... The search ended with "ke kou ke le." In Mandarin, this sounds very close to Coca-Cola and also has the positive meaning of "permitting the mouth to rejoice." Nice. Pepsi-Cola, by the way, translated its brand as "bai shi ke le," "everything that makes you happy."

In my book *Megatrends Asia*, published in 1995, I wrote that to work in China it is essential (and often required by law) to have a Chinese partner. To translate the meaning and emotion you want to be attributed to your brand's logo, it is indispensable to business survival. Drop even thinking of saving money here. Coca-Cola hired Alan Chan, one of the few established Hong Kong–trained graphic designers to refashion the brand name with artwork. Coca-Cola's first Chinese script logo since 1979 was launched in February 2003.

Despite all the problems, many U.S. companies are more deeply involved with China than most people realize. PepsiCo, for example, is China's largest private-sector potato farmer. Pepsi initially got involved in potato farming to increase quality and supply for its potato chip business in China. Not any potato will do it. They have to be grown in an ideal shape, ideally as round and large as a baseball, harvested at a time to meet the exact schedule to supply the purchase, and handled with the care of an egg. Not an easy task with China's traditional crude farming and distribution methods.

McDonald's has been working with suppliers to increase quality for years, but leasing farmland in Inner Mongolia and in Beihai, the southern part of the Guangxi province, to become farmers is a new experience for Pepsi. "PepsiCo is not a farming company," says Bob Shi, operating director for Pepsi's snack business in China. "But to build a market we had to take extra steps like this."

Pepsi has decided to build a global presence by dominating snacks rather than go head-to-head against Coke's soft drink dominance, although you wouldn't know that walking down Shanghai's greatest shopping boulevard, the two-mile Nanjing Road, where Pepsi signs outnumber Coke signs at least 4 to 1.

THE DRAGON DRESSES UP

Many years ago, when *Megatrends* was published in China, I was invited to lecture at Shanghai University. I remember it as an event where I was welcomed with great exuberance. I walked up a central aisle as the students rhythmically clapped me in. That never happened to me in the States.

My speech was probably quite emotional, and we had a great time through a long question-and-answer session. Again I got confirmation that Chinese students are among the most engaging and smartest of the world. After my speech, the students started to line up to get their copies of my book signed. In this quite pleasant process of signing, I slowly became aware that almost every book had a different cover, and only a few had the one my publisher had sent to me. It was clear: The knockoffs outnumbered the real ones by quite a margin.

I remembered this experience when I read about Giorgio Armani in China. It made me understand how he felt when he was offered a fake Armani watch for $22 in Shanghai. "It was an identical copy of an Emporio Armani watch," he said. The difference between Armani watches and books in China is that you don't expect to make money on the books, but you do worry about how faithful they are to the original manuscript.

As Armani put it, "fakes could cause some confusion among customers. On the other hand, it's flattering to be copied. If you are copied, you are doing the right thing." He definitely seemed much less amused when he said he knew of some stores in China—Giorgio Armani, Armani Fiori, Emporio Armani—that "have nothing to do with me."

It is predictably embedded in the present that once the Chinese experience copying of their own brands, they will blow the whistle on copying, just as Taiwan did.

Despite the bumpy ride you have to expect between the arithmetic of its population and its incredible rush to modernity, China is the magnet of opportunity. For local fashion pirates and for the world's biggest fashion brands, China's market is the new opium, and the big names are falling all over each other to build a business there. Within all the rapidly growing metropolises throughout China, Shanghai offers some of the greatest locations.

One of our favorites in Shanghai is Zhongshan Dong Lu along the mighty Huangpo River, well known as "the Bund." During Shanghai's heyday in the early twentieth century, the Bund was occupied by European, American, and Japanese banks, trading houses, consulates, and hotels, before the Communists took over and used them for their purposes.

In 2002, before my wife Doris and I went on our trip to Tsingtao with Fred and Ivy Li, we met in Shanghai with them to do some business. Our Chinese friends told us about a new place we absolutely shouldn't miss: M on the Bund. An Australian chef, Michelle Garnaut, had been taken by the dynamics of the Bund. She rented the top floor and terrace of a 1920s building and despite many and usual problems with local authorities, created a wonderful restaurant. She was, in a way, a pioneer, who had been warned that Shanghai was not ready for fine dining. Wrong—it was, and on a warm early summer night in May 2000 we sat on the terrace, eating crispy suckling pig, still our favorite dish there, and drank Australian wine. (China is still in a rather primitive stage of wine production.) Six floors beneath the restaurant, streams of red and white light trace the cars as they moved along the eight-lane boulevard. Across the river, the lights of the rapidly growing Pudong district were flickering through the darkness.

Shortly after M on the Bund had conquered the fine-dining crowd, it was clear that Shanghai was ready not only for sophisticated dining but for fine art and designer fabrics as well. Now, when we are in Shanghai, we not only go to the Bund to have our already traditional dinner at M but we also walk around in the art galleries that have opened in the building next door, have lunch in one of the three spectacular restaurants, and visit the new Armani flagship store, the beginning of a plan to add another 30 stores in China by 2008.

Armani is not the only one. Most big names are opening branches. The most noticeable successes include Alfred Dunhill and Louis Vuitton, which will add four stores this year for a total of it 13. LVMH (Möet Hennessy Louis Vuitton) says customers in mainland China are now ranked number four globally in their annual sales. Prada will have 25 stores in the mainland by the end of 2006. Zegna has 42 existing stores in more than a dozen cities. When I bought a Zegna suit in Beijing a few years ago, it was no more special than buying green tea in Milan.

Every big watch company is exploring China's potential. The Richemont Luxury Group (which includes Cartier) has been very active, as has the Swatch Group. Nick Hayek (son of founder Nicolas Hayek), the CEO of Swatch, is planning to open between 500 and 1,000 stores in China by 2010, when he expects to be selling up to 6 million watches a year. About 20 percent of his stores will be located in Shanghai.

China's luxury-goods market is growing 20 to 30 percent a year, the highest growth rate for the market in the world. Luxury goods is still a rare field in China, but Chinese competition is on the horizon. It will be like Japan, which supplied textiles for decades to the fashion industry before discovering the tremendous added value of shaping that material into high fashion. China's next.

KARL LI GA FEL ON THE WAY?

In October 2004 Naomi Campbell gave a speech on career development at Shanghai University. She was made guest artistic director of the modeling school affiliated with the university. Shen Zhiwen, vice dean of the modeling school, said other international supermodels will be invited to give classes and provide instructions in the future. "China is the world's biggest garment manufacturer and consumer of clothes," he said. "There are more than 50,000 garment companies, but only about 10,000 professional models in the country, which leaves a big gap. The international catwalk needs more Asian models, especially Chinese models, because European fashion houses want to break into this huge market."

Shanghai opened the doors to its first fashion week, organized by the city government, in October 2003, launching an international fashion festival in the spring and a fashion week in the autumn, seen as an important step in catching up with other international fashion centers.

In Beijing hundreds of fashion aficionados, businesspeople, models, photographers, and journalists from all over the world converged to take part in China Fashion Week in spring 2004. The week's catwalk shows offered an ample opportunity for designers from both home and abroad to show off their autumn and winter collections. One year later at Beijing Fashion Week, 33 new designers from 19 countries were showing their collections under the theme "My Vision of China." The show hosted the 12th International Fashion Contest for Young Designers, the Hempel Award, giving new talent a chance to shine.

Young designers from abroad, especially from the United States and Europe, are parachuting into Beijing and Shanghai for design opportunities not available at home. Vera Wang, who is known for her American bridal empire, has become the first designer with Chinese roots to be globally recognized. "We have to work more on both fabric and design to win consumers and try to enter the world market soon," she said.

Wang achieved recognition her parents could never have imagined. Her mother and father, the son of the war minister under Chiang Kai-shek, left their native China in 1947 to start a new life in the U.S. Going back to her heritage, Wang has opened a bridal boutique, The Perfect Wedding, in Shanghai's Pudong Shangri-La Hotel. It offers the stylish, serene, softly colored outfits that have brought a new sophistication to the white wedding world.

Chinese are becoming aware of design as added value—for them. Today, China has more than 400 schools offering design courses that together graduate some 10,000 industrial designers a year, up from only about 1,500 in the year 2000. Design is now one of the most popular occupational preparations in Chinese universities. Hundreds of design consultancies are popping up, particularly in the three great cities: Shanghai, Beijing, and Guangzhou. Chinese companies are creating design departments within their companies with expanding staffs.

"Design is the way companies improve their competitiveness," says Yu Zida, the vice president who oversees designs at appliance maker Haier Group Co., the giant Chinese company that sells its products in 150 countries and was in the news for a time in 2005 as a possible buyer of Maytag. The Haier group has 120 industrial designers on its staff.

For seven years, Haier America has been operating in the States, where its biggest business is a line of stand-alone wine cellars in addition to small refrigerators and other appliances. Most people don't realize it is a Chinese company. The name was changed to *Haier* in the early 1990s when the company had a German partner.

Haier and Lenovo, the computer maker that bought IBM's PC division in May 2005, are now the two best known Chinese brands. Lenovo, always known for good design, has recently doubled its design team to 80 people and in 2006 won an Industrial Design Excellence Award for its new smart phone. Sony, Samsung, Motorola, Nokia, General Motors, and Volkswagen, among others, in trying to keep up with the Chinese, have opened design shops in China. Global companies think that as the Chinese get better at design, they need to develop products specifically for the Chinese psyche.

In China it is more important to appear rich than to be rich, and some products developed for the Chinese market don't travel very well. The Lenovo Group has had fabulous success with a mobile phone that holds a few drops of perfume, filling the room with sweet smells as the battery heats up. Volkswagen designers in Shanghai now use fake wood to dress up the inside of a car, a real yesterday look for most of us, but Chinese car buyers want it because it makes their cars stand out. "Understatement in China is a no-no," says Stefan Fritschi, chief designer at Volkswagen's Shanghai operation. "You want to impress your neighbor." That makes sense to me. I remember that when Chinese families first started to purchase refrigerators, they would set them up in the living room to impress visitors.

BRANSON'S VIRGIN SCHMOOZES THE DRAGON

Sir Richard Branson has big Chinese plans: "For any business these days, it is madness to not be involved in China." His man on the ground, David Baxby, head of Virgin Asia Pacific, in 2006 is relocating from Sydney to Shanghai.

Virgin Atlantic has offered daily London–Shanghai service since the end of 2005 and will soon add Beijing to that schedule. During 2006, it also doubled its London–Hong Kong service to twice daily. At the beginning of 2006, Asia provided 10 percent of Virgin's $20 billion revenue; Branson expects it to rise to 25 percent within five years.

Also in Virgin's plans is a launch of Virgin Mobile, a cell phone service, in China. "We think we can do something there. If we get it right, it could be an enormous market," Branson says, echoing a line often heard about China. Any mobile deal in China would be similar to Virgin's operations in the U.S. and Britain, with Virgin providing the brand, marketing, and promotional expertise. Whether or not Branson can establish a Chinese mobile phone business, Chinese consumers are about to see—and taste—a lot more of the Virgin brand over the coming months.

Sleeping in a Virgin's bed is now within reach. Branson is planning a chain of Virgin hotels and health clubs to open in 2008. Refreshing Virgin-branded soft drinks will be available as well.

At the end of 2005, Virgin opened one of its music megastores in Shanghai, an audacious move, considering China's slick counterfeiters. But Branson is nothing if not audacious: "Some people think we are mad because of the piracy problem, but my feeling is that we need to offer Chinese consumers a legitimate outlet." A Shanghainese friend said to us when we were there in November 2005: "Maybe we will really be able to buy decent music in Shanghai. Many think that the West has [only] a few artists: Elton John, Mariah Carey, the Vienna Boys' Choir, and Pavarotti." Thanks to Branson, a few thousand can now be added to the list.

Branson is not only widening assortments; Virgin is willing to open the door into other spheres. Just before Christmas 2005, he unveiled the launch site for Virgin Galactic, his space tourist enterprise. Don't expect me to tell you what Branson will conquer first, space or China.

THE RISE OF THE RICH

One important measure of the extent to which capitalism is progressing in China is the increase in the number of wealthy individuals. There are now 7 billionaires in China and more than 100 additional individuals with fortunes averaging half a billion U.S. dollars. Merrill Lynch recently estimated that the country now has 300,000 millionaires—a number that others have judged to be from "conservative" to "very conservative."

Economist Milton Friedman for years has talked about the choice between socialism and capitalism as a choice between economic fairness and economic freedom. Economic "fairness" is on the socialist side with its redistribution schemes. If you choose economic freedom, the capitalist side, you choose the freedom for people to make as much money as they can, which means that necessarily some people who are very talented at making money will make a lot of money. So by a kind of inverse measurement, we see that economically China, with its many millionaires and billionaires, has chosen economic freedom, which plays nicely into the long history of Chinese entrepreneurialism, practiced during the Communist years mostly by the Overseas Chinese. In watching the score of the game, do not neglect to gauge the economic strength of the Overseas Chinese.

CHINA'S FLAG BEARERS

If China is on the path to seek global supremacy, its sportsmen will be the first to arrive. Sport is "war minus the shooting," said George Orwell. As the nation-state recedes, sports fans become the new global flag bearers. We are moving into a great era of spectator sports. How do we know that this phenomenon will continue and grow? Because we have seen in the globalization process that the more economically interdependent people become, the more they assert their cultural/national identity. So does China. National sports teams give people a strong feeling of national community, something to identify with as they become increasingly dependent on the world economically.

The ground is prepared: In the 2004 Athens Olympic games, China won 32 gold medals compared with 35 gold medals for the Americans.

Nobody should be surprised if China wins more gold than anybody in 2008 in Beijing.

Basketball is not new to the Chinese. It was introduced to them by Christian missionaries shortly after it was invented by James Naismith in Massachusetts in 1896. It is hugely popular in China, where it has been played for 100 years. In the 2004 Olympics, China came in eighth in basketball. But the lead that the United States still has might be used up quickly. In 10 years, according to Xia Song, a sports marketing executive and game announcer, "Chinese basketball can be one of the dominating forces in the world."

The first NBA game played in China took place in 2005 in Shanghai: the Sacramento Kings against the Houston Rockets, with the outsourced Yao Ming, the tallest basketball player in the world, standing out as a great link between China and America.

Since 1997, China has had a pro basketball league of 14 teams. As part of a new improvement program to raise the skill level of its players, the Chinese Basketball Association held tryouts in the United States in 2005 and is importing—outsourcing—2 American players for each of the 14 teams. The Chinese teams are allowed to pay the American players up to $30,000 a month—more than that offered by many European teams.

In the other direction, Yao Ming is China's most famous and best paid flag bearer. There is talk in the basketball world of a new Ming dynasty. Almost as popular as Yao Ming in China is Allen Iverson, who at this writing, is the leading scorer in the NBA. At 6'1" versus Ming's 7'5", Iverson is of a more "natural size," a superstar whom more Chinese can relate to.

The competition to bring world sports events to China's cities—another step in the long march from gray Communism to the colorful culture of sports and celebrity—brought the biggest sports businesses to China, in 2004 to Shanghai, with the running of the first Formula One race in the country's history. The stadium for the event cost $325 million to build and has 150,000 seats, all of which were sold months in advance the first year it took place. Michael Schumacher, who, by the way finished in eleventh place, said it was "the best racetrack I have ever seen." To join the Formula One spectator circuit is not the only goal in China. Chinese *drivers* are now working their way up through the ranks in Formula 3.

All this, of course, is part of a larger plan by China to develop as a power in world sports, and "another sign of Shanghai's relentless effort to stage more prestige events than anywhere else on the planet." China will dominate global sports long before it dominates the global economy—if it ever does.

CARS IN CHINA

A farmer on a cart pulled by a water buffalo is an iconic picture of China, as are the hordes of bicycle riders in the cities. But these pictures are rapidly becoming a part of China's past. With more and more people living in sprawling cities, distances can no longer be managed by foot or bicycle. A growing affluence and the relative low price for a car, the cheapest around $3,600, make cars not only desirable but also affordable to more and more Chinese people. In 2005 China exported more vehicles than it imported—a milestone in the country's ambition to be a leading player in the global car domain. The Chinese government expects that China will make and sell 6.4 million passenger cars, vans, buses, and trucks in 2006, surpassing Germany as the third largest vehicle maker after the United States and Japan.

A milestone, yes, but it's only a beginning. China's vehicle exports mostly go to countries in the Middle East and Africa. Many of the cars coming into China are luxury brands from the West.

DaimlerChrysler announced in 2005 that the company wanted to export small cars from China to the United States. With one of its Chinese joint-venture partners, the company is looking to build a factory for the exports. In addition to engaging in joint ventures, Chinese auto companies are gearing up to sell cars in the United States and Europe themselves. Bob Lutz, the high-profile vice chairman of General Motors, says that "we're rapidly approaching the point" where at least one of China's homegrown automakers will be successfully exporting around the globe within the next five years.

In the best assessment I have heard about the quality of Chinese cars, J. M. Noh, the president of Beijing Hyundai and an automotive-quality expert, says that "in terms of quality, the cars from China and the cars from Korea are the same." And these cars are being built by Chinese industrial workers at $2 an hour or less, including benefits—one of the lowest pay rates in the world. The hourly cost of wages and benefits in the big three auto-making countries—Germany, Japan, and the United States—are $49.60, $40.96, and $36.55 (given here in U.S. equivalents), respectively. Honda has just finished a factory in Guangzhou for exporting cars to Europe beginning in 2006. The Chinese carmaker Chery Automobile plans to begin shipping cars to the United States in 2007.

The growth that is dearly embraced in an increasingly affluent consumer market creates a lot of hand-wringing about environmental consequences. While automobile makers welcome every new motorist, they are faced with the concern about emissions being released into the atmosphere by cars those motorists drive. In turn, the complaint in China and other parts of Asia is that the West, having gotten its cars and having polluted the atmosphere, wants to stop others in the name of the good of the planet. In this regard, the West is seen as not having the moral standing to ask for restraint on the part of the East. Several years ago, I suggested a solution: The West should pledge to reduce the number of its cars to match the number of new cars added in China. For each car added, the West subtracts one car. As you might imagine, I got very few takers—in fact, none. So how serious is the West?

TOURISM

As incomes rises and the freedom of travel increases, more Chinese will travel abroad. According to the Word Tourism Organization, there will be more tourists from China—115 million every year—than from any other country, including the United States, by 2020. Travel barriers for the Chinese are coming down dramatically. Already in the year 2004 there were 29 million departures from the mainland, although about 75 percent of outbound tourism was to Hong Kong and Macau.

FILMMAKING

Hollywood has seen the future, and it is China. Walt Disney Pictures is making big plans for picture making in China, including a martial-arts remake of *Snow White*, with monks replacing the dwarfs. *Crouching Tiger, Hidden Dragon* got everybody's attention, and now Western studios are jumping at the chance to make China into a major film production base for both external and internal consumption, for both English- and Chinese-language films.

"China is going to grow, so a lot of companies want to come here and produce films," says Li Chow, general manager of Columbia TriStar Film Distributors. "Chinese films have done well internationally, ever since 'Crouching Tiger' came out [in 2000]." Hollywood is totally aware that China now has world-class film directors who are breaking box office records in China and are increasingly successful in selling well in other countries.

Zhang Yimou's last two movies, *Hero* and *House of Flying Daggers* (a stunningly beautiful action film), have together grossed almost $200 million outside China. The biggest Chinese-language success so far, Ang Lee's *Crouching Tiger, Hidden Dragon*, has become the highest-grossing foreign-language film in the United States, earning at last report $128 million. Yes, Ang Lee won an Oscar in 2006 for best director for his film *Brokeback Mountain*. This was a case where a Chinese director handled a U.S. film. In the other direction, Miramax's *Kill Bill* films were largely shot in China.

There is now a Chinese-language version of *Variety* published in China, and the *Hollywood Reporter* has opened a Beijing bureau. "Why am I here?" asks the new bureau chief, Jonathan Landreth. "Because everyone else in Hollywood is."

CHINA VERSUS INDIA

Almost two of every three people in the world are either Chinese or Indian. The way it is now set up by the media, there is a great race under way between China and India to see which will be the leading power by the year 2100. But so far, China has outperformed India on every measure of economic growth and poverty reduction.

Whenever China is in someone's mind, a concern about India is not far away. When I think of India, I somehow then think of Brazil. For so many years, we have talked about Brazil's great potential, great future, but we have never seen it.

When China started its reforms in 1978, it was poorer than India. India has recently had a new economic vitality, but politics has now brought further economic reform to a halt. When I return to Shanghai, even after an absence of only a year, it is a dazzling experience: new buildings; old buildings torn down and replaced by building cranes and shortly new buildings, most likely skyscrapers; new restaurants, hotels, and shops opened. When I return to Delhi or Mumbai, however, it always seems the same as it was before, including the extraordinary number of beggars at every turn.

India has such great potential, such a great future—a great future that fails to come? We are still waiting. India's economy is now growing at 8.4 percent. But where are the real economic reforms?

I was last in India during July of 2006 when I keynoted IBM India's "Forum 2006" at meetings in Mumbai and New Delhi. The IT sector, with all its outsourcing, is of course the big success story in India. It is in fact the only big success story in India. This is somehow taken as the great awakening of the Indian economy. But IT is only one sector, and you can't build an economy on the IT sector alone (and already Indian companies are outsourcing IT work to China). The IT sector is succeeding in spite of the government. The entrepreneurs driving the sector have had to outwit the heavy hand of the hangover socialism of the government. The tentative economic reform efforts of the early 1990s have ground to a halt, and the overall economy will not develop until real economic reforms are put in place—not likely, as it appears today.

The recently announced merger in the Steel Domain is instructive. India's biggest steel company, Mittal, and Europe's Arcelor are merging to become the world's largest steel group. So now we will have an Indian steel company that will be by far the biggest in the world, producing 10 percent of the world's output, three times bigger than its nearest rival, Nippon Steel, with less than 3 percent of the market.

India will have the biggest steel company but produce no steel in India—no contribution to India's GDP. Mittal Arcelor will only contribute to the GDPs of the many countries where it actually produces steel. And Lakshmi Mittal, the entrepreneur behind it all, lives in London. Welcome to the new Economic Domain world.

DIFFERENT PUZZLE PIECES

India has a good banking system, a 150-year-old stock market, many well-managed companies, and a huge, well-educated middle class. But politics and nationalism have smothered economic reforms for years. The new prime minister, Manmohan Singh, is a great economic reformer, but his first speeches were from a socialist platform rather than an economic reform platform.

China took in $55 billion in foreign direct investment in 2004, compared with $5.3 billion for India. That almost 10-to-1 ratio favoring China over India has been the same for almost a decade.

China should also be seen as an economic leader over India because of its great infrastructure advantage. China has almost 20,000 miles of expressways, 10 times that of India, and 7 times as many conventional land telephones and mobile telephones per 1,000 people, though what infrastructure India does have today is a great improvement on the past. I remember giving a lecture in Mumbai (Bombay at the time) in the early 1990s, and after learning that there was a ratio of 1 telephone operator for every 12 customers, I declared, to boisterous laughter, "That's not a telephone system. That's an employment program!"

In dealing with India's poor infrastructure, Simon Long of the *Economist* writes that "even Bangalore, epicenter of India's IT industry, suffers from traffic jams, overflowing hotels, power cuts and an inadequate airport. It risks throwing away its great advantage: that it has attracted a critical mass of the world's high-technology firms to what could be a self-sustaining cluster and boomtown." India, despite all the information technology you read about, has one of the lowest personal computer penetration rates in the world, far behind that of China.

Judgments about India versus China suffer from the excitement of witnessing the growth in China. India's GDP grew by an average of 5.6 percent a year in the 1980s and by 5.8 percent a year from 1991 to 2003.

China's grew by 9.3 percent in the 1980s and by an average of 9.7 percent between 1991 and 2003. In 2004 China overtook Japan to become the world's third largest trader behind Germany and the United States.

The exchange rate of China's currency, the renminbi (often referred to by its old name, the yuan), is a huge issue in the global economy. When was the last time you heard anyone get excited about the rupee?

Long says that perhaps the most telling single statistic about the difference between India and China is female illiteracy rates. He quotes the World Bank as reporting that 87 percent of adult Chinese women are literate, compared with India's 45 percent. Adult literacy is 91 percent in China versus 57 percent in India.

When Deng Xiaoping visited the Ford plant near Atlanta in 1979, he said: "We want to learn from you." Deng and his successors learned fast. China, well on its way to becoming the world's largest carmaker, is today on the path to becoming the industrial power Deng envisioned.

Remembering that former president Jiang Zemin told me that during the Cultural Revolution he was an assembly-line worker in an auto plant in Shanghai, I wonder what he thinks of all this now.

Applying Mindsets

MINDSET #8   "Things we expect to happen always happen more slowly": We are fast in speculation, but things unfold at their own speed. Hold back when China is said to take over the world next Tuesday morning. For thousands of years, China has thought in Dynasty time frames. This new period, the new dynasty of free market, globalization and decentralization, has just begun, and will take many years to play out.

MINDSET #5   "See the future as a picture puzzle": China itself is very similar to a picture puzzle, decentralizing as no other country in the world. Only by observing what is going on in its parts, the regions and cities in fierce competition with the world, will we understand the place China will have in the big picture of the world.

MINDSET #9   "You don't get results by solving problems, but by exploiting opportunities." China has become a nation of opportunity seekers, trusting that any problem can be solved along the way. The competitive advantage of today's China, as a nation and as a people, is the willingness to adapt to what is necessary and beneficial. No other country is so massively adapting to change as China; few peoples are so entrepreneurial and aggressively ambitious.

Chapter 4

EUROPE

Mutually Assured Decline

The "Statue of Europe" has two hearts and 25 mindsets. The 25 country mindsets are stirring a mixture with ingredients that do not blend: tradition, ambition, welfare, and economic leadership. Her two hearts beat in different rhythm, one for economic supremacy and one for social welfare. Proud and ambitious, each one wants to be right. But to reach either goal, they have to compromise, and neither side is willing to do so. My experience makes me believe that Europe is much more likely to become a history theme park for well-off Americans and Asians than the world's most economically dynamic region as it has proclaimed it wants to be. Economically, Europe is on the path of Mutually Assured Decline.

EUROPE'S PROUD AMBITION

Over the many years since I started to give speeches in various places in the world, I have visited Europe many times and even owned an apartment in Paris for a while. My experience gave me a good understanding of European culture and thinking. But there was much more to learn, something I became aware of when my connection with Europe and especially Vienna began to tighten. I married my German-language publisher, certainly giving new meaning to author–publisher relationships.

The year 2000 was not a bad time to move to Vienna. I could experience firsthand how Europe as a union would change. The excitement about the European Union and the upcoming reality of the euro was in the air. Austria, for centuries the seat of the Habsburg empire, would once more be in the geographical center. Vienna would again be the bridge between the east and the west. Optimism was the prevailing mood.

The path to the European Union had started in 1951 with the European Coal and Steel Community agreement among France, Germany, Italy, Luxemburg, the Netherlands, and Belgium. At some point along the way, Europeans started to trust the process of becoming the European Union. In looking back on a history of war, the miracle of course was the rejection of a 700-year tradition of settling problems through military power. No more killing each other. Problems would be dealt with solely through diplomacy, international law, negotiations, and multilateralism. Forming the European Union was a sea change, the greatest geopolitical change in the history of the world.

The Lisbon Accord in 2000 then made it clear that the goal was "to make the European Union the most competitive and dynamic knowledge-based economy in the world." In other words, to challenge the primacy of the United States. This is where the miracle has come to a halt. Maybe the miracle would have continued, were there not the two hearts that cannot find a common rhythm. Without question, the European Union could well agree with Goethe's Faust: "Two souls, alas! Are lodg'd within my breast."

One heart, the socialist heart and soul, is beating in the rhythm of regulation, centralization, strict labor laws, and a redistribution of wealth. The other heart, the conservative one (or the liberal one, as Europeans would say), is beating in the rhythm of a free market, decentralization, and competition and support of entrepreneurship. The socialist–green coalition held the flag of economic supremacy high, but despite the continuous lip service given to necessary economic reform, with Germany and France forming the leading voice, Europe held—and holds on to—the old mindset of the "European social model," which is getting harder and harder to sustain. Europe is increasingly losing ground in trying to become the world's economic driver, because it dearly embraces what one of its famous sons, Sigmund Freud, wrote: "It is easier to suffer than to act." The dilemma is not eased by the fact that both sides insist on being right. Endless political discussions are not about how to exploit the new opportunities of housing economic domains, support entrepreneurship, and attract the talented people in the world to come and strengthen Europe, but about how to maintain the welfare state that now yields slow growth and unemployment. Unless serious reforms are put in place, the most likely outcome is that instead of leading a knowledge-driven global economy, Europe will become a place where people come to witness a glorious past.

If you want to make your own judgment and evaluate what is going on in Europe and where it is heading, you need to check the score of the game. What is embedded in Europe's present?

IN SEARCH OF THE "TRUE EUROPEAN SOCIAL MODEL"

The discussion throughout Europe is as necessary as it is difficult. It would seem enough to have the two hearts, one beating for and one beating against the priority of welfare, but in addition, the 25 mindsets of the 25 European Union countries are pulling in different directions. Each one is looking for arguments supporting its own opinions and needs, such as the impossibility of financing the system, the economic slowdown it causes versus the perceived moral duty of redistributing wealth, and the unfairness of a free market. It is not that the supporters of the social model do not see the economic problems, or that the other side does not see the benefits. What is missing is a uniting compromise between two views of life—humanistic ideals and economic realities—in an emotional quarrel.

Anna Diamantopoulou, a former European Union commissioner, put it this way: "For some in the EU, the expression 'European Social Model' evokes warm feelings of social justice and solidarity. For others, it just raises the blood pressure."

What are the characteristics of the European social model? Here the trouble begins. There is no such thing as one consistent European model, but there are many different groups, all with different focuses. The *Economist* came up with an interesting comparison: "There are almost as many forms of social policy as there are national drinks." Cheers! And at the end, Europe gets drunk on every one. However, the 25 mindsets can be arranged into four groups with certain similar characteristics.

THE CONTINENTAL SYSTEM

The continental system was founded by Otto von Bismarck, the first German chancellor (1871–1890). He was the first statesman in Europe who devised a comprehensive form of social security offering workers insurance against accident, sickness, and old age. This entitled a person to get a retirement payment beginning at age 70. The average life expectancy at the time was 40 to 50 years. Today life expectancy is about 78 years. Bismarck's system was intended not to force redistribution but to force citizens to insure themselves. Today the system is in place in Germany, France, Austria, Belgium, and other countries nearby. Because of increasing pressure mainly from the aging population, the insurance character of this system cannot be sustained.

If Johannes Schmidt loses his job in Germany, he will receive 60 percent of his net wages for 12 to 36 months, not counting other government help, including training and support in finding a new job.

The Mediterranean System

Mediterranean countries such as Italy, Spain, and Greece have strict laws protecting jobs but modest financial unemployment support. Social security is still considered a responsibility of the family; the whole system relies on internal family support. Not everybody is covered by the system.

If Giacomo Esmido works in Italy and loses his job, he will receive up to 40 percent of his last pay rate for up to 180 days.

The Scandinavian System

In Scandinavian countries, we find the strongest social protection. The system is mainly financed by taxes. They have few laws securing jobs but high support for unemployed people and focus on helping people to find work. Social welfare is considered a right, and strong governmental protection is offered. A high redistribution of incomes takes place. If you want to be unemployed in Europe, you should definitely go to a Scandinavian country.

If Jens Sørensen loses his job in Denmark, he will be better off. As a childless 40-year-old, he will get 90 percent of his gross wages for no less than four years while out of work. Remarkable. But such a system has its price. In Germany the tax revenue comes to 45 percent, in Denmark to 49.6 percent, and in Sweden to 50.7 percent. That's the other side of the coin.

The Irish and U.K. System

Ireland and the United Kingdom have weaker job protection and few redistributive taxes, but support to find a job is offered. Also, generous unemployment benefits are available, which still separates them from the United States. The focus is not on reducing income equality but on fighting poverty.

If John Smith loses his job in England, he receives support payments of $400 per month, for an indefinite period, with additional support the first 182 days.

Vladimir Spidla, prime minister of the Czech Republic and also one of the 25 European Union commissioners, names the values of the social model: "Although there are national variants, the underlying principle of the European social model is that all inhabitants are covered by basic social insurance and nobody is abandoned. Its purpose is to provide people with support in difficult situations, including illness and accidents, natural disasters, and changes in the economy and living conditions that prevent proper education and access to the labour market." There is nothing wrong with that wish. But as long as Europe does not discover an oil field of at least Saudi Arabian dimensions, I see no way to afford it.

The supporters of social welfare often emphasize the difference with the United States as a threat to Europe and as a strong argument against the United States. In a vacation paradise of *at least* four weeks, only two weeks a year is almost unthinkable. In fact, an average European works only 70 percent of the time that an American does. But to cut down on the social comfort of more than a century would be as consequential as slaughtering a holy cow in India.

The *Wall Street Journal Europe* points out that European politicians are more interested in the theoretical value of the system than in the practical relevance. "And in theory, the European Model is very significant, because it helps in setting Europe apart from the rival: America." The flag of humanism has to flutter from Europe.

Experiencing Europe as a resident, I agree with what Joshua Livestro wrote in the *Wall Street Journal Europe*: "The European Model, you see, is not just an economic model. It's a mindset, a way of life, a worldview if you like."

It's a lifestyle with some comfort. I well remember the days my phone calls to my Viennese publisher accumulated and how astonished I was when on so many days she would say: "We don't work tomorrow; it's a holiday." Europe today is stirring a mixture with ingredients that do not blend: tradition, ambition, welfare, and economic leadership. All of it is topped by the politics of envy. It might be popular, but newspaper articles and political statements pointing at "unfair wealth" and crying for active redistribution of wealth will not get Europe one step further toward fulfilling its ambitious economic dreams.

Austria's chancellor, Wolfgang Schüssel, had a gestalt into reality when he modified an old Austrian saying: "We have to switch from an island of the blessed to an island of the diligent."

FROM COMMUNIST WATCHTOWERS TO CAPITALIST PLANTS

On my first trip to Vienna, my hosts took me to a little lakeside village called Mörbisch, well known for its many storks and their nesting sites. Through the middle of the lake, they let me know, the Austria–Hungary borderline was drawn. They took me to the land-bound part of the border. As we walked along the field road toward the border, signs with skull and crossbones came into view, ominously warning that stepping closer would cause serious trouble. I felt uncomfortable when I saw the wooden Hungarian watchtowers lined up about every 200 meters. On a little platform on top of the towers, border guards were patrolling with their guns, ready to shoot at anybody who attempted to cross. That was only 30 years ago.

Today, Hungary is part of the European Union, and when we visit its beautiful capital, Budapest, we simply drive through the border with easygoing passport control. In the late 1980s Hungary was the first to make a hole in the Iron Curtain by letting East Germans cross the border to Austria, and it was the first to open for foreign investors in the 1990s.

In Europe's most recent enlargement, Hungary, Slovakia, Estonia, Slovenia, the Czech Republic, and other small countries formerly in the Soviet orbit, have moved to higher ground. They are now more free market, business friendly, and economically free than stuck-in-the-mud France and Germany. These new members of the European Union are competing with low tax and high support to host as many companies as possible. In an economic domain world, it doesn't matter where you are based or where your economic activities are. In just the last two years, plants and people in Slovakia have become important parts of the automobile domain. Slovakia has a flat personal income tax of 19 percent that also applies to companies. All those former poor countries are doing well economically because of their participation in and attractiveness to economic domains.

IMMIGRATION: CHANCE OR FRIGHT?

Not only do I come from a nation of immigrants but I also am a first-generation American. My father was born in Scotland; my mother was born in Denmark. As my wife Doris says, I have come back to my European roots. But even with this background and after living in Europe for six years, I do not understand the xenophobic thinking that I experience here. A few years ago, I was invited to speak at a conference of the Austrian government on the subject of sustainability. There was a lot of talk about global warming, natural resources, genetically modified food, and so on, but not a single word on *the sustainability of Europe's population*.

Why would you resist having immigrants when you need workers because of a declining population, when you need talent, when you need energetic populations? And at a conference on sustainability, why never mention population sustainability?

But many in Europe say that immigrants only seek the benefits of the European welfare model, while the unions complain that immigrants only take their members' jobs. In a dynamic economy, jobs are lost all the time; the churning of jobs is the accommodation to changing times.

The low birth rate in Europe, plus the general unwillingness to welcome immigrants, will soon catch up with Europe. Europe's birth rate is only 1.4—that is, women in Europe now give birth to an average of 1.4 children. It takes an average of 2.1 children just to sustain a population at a constant level. If the birth rate in Europe does not increase (and it seems unlikely that it will) and if Europe continues to limit immigration, then in just two generations, Europe will have half the population it has today.

As the *Economist* has observed, "The long-term logic of demography seems likely to entrench America's power and to widen existing transatlantic rifts," providing a stark "contrast between youthful, exuberant, multicolored America and aging, decrepit, inward-looking Europe." Tough words, but that's the score of the game.

Despite the ideals and wishful thinking, there is no way of denying the necessity of a change in the European mindset. The government's job is not to provide full employment and wealth. And it can't provide them anyway. The aging population will make the system unaffordable. It is already unaffordable. Nevertheless, the solution of allowing immigration to make up for the much-too-low birth rate is widely condemned and is a popular political threat used to make people fear job loss.

America has devised a way of enriching its talent pool. It takes in more than 1 million legal immigrants each year, more than all other countries put together—and that has been the case since 1976. Who are these 1 million people? For the most part, they are talented, ambitious, smart people who want to realize their dreams in a place where they have the freedom to do so. One million a year added to America's talent pool. Because brains and talent are randomly distributed, think what the domestic birth rate would have to be to get the same result.

My experience is that change comes when there is a confluence of changing values and economic necessity. The economic necessity seems clear in Europe, but Europe does not seem to change even when benefits are obvious.

The condition that will create a competitive economy is a nourishing environment for entrepreneurialism. In Europe the thinking is that the government should ensure that no one is fired and that new jobs can be created by funding government projects. Japan has done this for years and now has more bridges per capita over almost imperceptible streams of water than any other country in the world.

Only entrepreneurs, bottom-up, can create new companies and real jobs. This should be done by eliminating all the bureaucratic impediments that have been placed in the path of creating new companies, and creating such incentives as a waiving of taxes for the first years of a company's life. The larger issue here, I think, is that it is not understood that a company needs to make a profit to survive. A lot of Europeans seem to think that companies can choose between making a profit and keeping more workers on the job even when losing money.

Sure, the European Union now has goals for emission reductions, renewable energies, biodiversity, and social inclusion. But they have little impact on economic growth, and in some cases slow it down. Economic reforms are not moving at all. The social model enjoys such popularity that hardly any politicians speak against it out loud. Some do say it should be reduced, or made more efficient, but the model itself is hardly questioned.

Economic reform is about improving an economy's productive capacity. Period. All talk abut economic reform should be judged against this standard. As it stands now, Europe for the most part is hostile to entrepreneurs, just when it needs them most.

WHO HAS THE RIGHT TO RUN THE WORLD?

Europe without doubt understands itself as the flag bearer of humanism that emerged as the characterizing spirit during the fourteenth through sixteenth centuries in Italy, in the cultural and historical epoch of the Renaissance. Philosophically, humanism is defined as any outlook or way of life centered on human need and interest. That, in many ways, is seen as the European inheritance and ideal.

Clearly, Europe feels superior to the picture that is painted of America, led by the French, who defined President George W. Bush's America as a triple-headed hydra breathing flames of imperialism, neoconservatism, and fundamentalism. The rivalry between Europe, which feels an intellectual superiority and therefore claims the moral right for economic and military supremacy, and the United States, which is much more relaxed about its own way of life and certainly has economic and military supremacy, is inevitable.

HAYEK: A EUROPEAN EMIGRANT'S DESTINY

In 1950 Austrian economist Friedrich Hayek boarded a ship in London to accept a position at the University of Chicago. He had been widely ignored in Europe but was being embraced in the United States. Trained as an economist in the twentieth century, he was on his way to becoming the most important social philosopher of the twenty-first century. Stanford economist Thomas Sowell calls Hayek "the central pioneering figure in changing the course of thought in the twentieth century."

Why did Europe ignore him? As opposed to his contemporaries, Hayek did not believe that the best way forward was through central planning. Centralized systems may look attractive on paper, he argued, but they suffer from a basic and incurable ailment: the "division of knowledge" problem. In order to know where resources should be directed, the central planner needs to know both what goods people want to buy and how they can most cheaply be produced.

But this knowledge is held in the minds of individual consumers and businesspeople, not in a government planning agency, and the only practical way for customers and companies to relay this knowledge to one another, Hayek argued, is through a system of market-determined prices. The great lesson that Hayek taught us is that central planning and individual freedom can't coexist. Fundamentally, Hayek wanted *human societies to develop spontaneously in freedom*—the best way to cope with the unknown.

However we construct our societal arrangements in this world, I would like to tell you why, in the shift from the declining welfare state to self-organizing market economies, Friedrich Hayek is becoming so important and why Europe should listen to him.

1. He is our greatest philosopher of liberty. For Hayek, the individual is paramount. He concluded his famous book, *The Road to Serfdom*, saying "a policy of freedom for the individual is the only truly progressive policy." This should be a guiding light for Europe.

2. He stood for the proposition that free-market capitalism is the only practical way to organize a modern society and that the key to economic growth is "knowledge." He articulated these views in a very modern way. He saw capitalism as a spontaneous

information-processing machine. He spoke of this processing as a "telecommunications system"—a staggering insight at the time. These views have now become almost universally accepted, but are still fighting for acceptance in Europe.

3. He stuck to his guns. *The Road to Serfdom* (1944) was an attack on the welfare states that were being built in wartime Europe. He was a voice in the wilderness, deeply unfashionable. In England he was called a "magnificent dinosaur." *The Road to Serfdom* was addressed mostly to Europe, which paid the least attention to it. The resistance to change and the need to be right is still blocking the way.

Hayek received some international approval when he received the Nobel Prize in 1974 together with Gunnar Myrdal. At the time, Europe was still denying his importance, and many said in effect, "Wonderful about Myrdal, too bad about Hayek." Myrdal later said that he wouldn't have accepted the award if he had known he would have to share it with Hayek.

In 1989, three years before he died, Hayek watched the fall of the Berlin Wall, the embodiment of the failure and breakdown of the socialist republic, on television from his home in Freiburg, Germany. It must have been a sweet moment!

The nation that has embraced Hayek the most is now the nation Europe questions the most—America.

HATED, ADMIRED, ENVIED, COPIED

Sometimes the most uniting thing is to share an enemy.

America, the thorn in Europe's political flesh, was probably the most bonding catalyst among Jacques Chirac, Gerhard Schroeder, and other politicians who played the first fiddles in the orchestra of the America-bashers, welcoming guest stars such as filmmaker Michael Moore with pleasure. I felt that there was a hidden idea that if the world needs to be tuned to democracy, the music has to be conducted by Europe.

Europe of course has a long history of colonizing, in some cases well into the twentieth century. "Europe saw the world as a self-service store," as Radio Bremen, a German radio station, once said. Does Europe challenge America's effort to democratize the world because of its own history? Partly.

Michael Ignatieff, director of the Carr Center for Human Rights Policy at the Kennedy School of Government at Harvard, describes America as an "empire lite":

America's empire is not like empires of times past, built on colonies, conquest and the white man's burden. The 21st century emporium is a new invention in the annals of political science, an empire lite, a global hegemony whose grace notes are free markets, human rights and democracy, enforced by the most awesome military power the world has ever known.... The President's National Security Strategy, announced in September, commits America to lead other nations toward the single sustainable model for national success by which he meant free markets and liberal democracy.

Clearly, a strong mindset.

...Sept. 11th changed everyone, including a laconic and anti-rhetorical President. His messianic note may be new to him, but it is not new to his office. It has been present in the American vocabulary at least since Woodrow Wilson went to Versailles in 1919 and told the world that he wanted to make it safe for democracy.

America was an empire long before George W. Bush came along. It may surprise you that America has had 10 presidents who were former generals. Except for Vietnam, America has never lost a war, a record unprecedented in modern times. Not since the Roman Empire has one country so dominated the world. For a good historical account of America's march to empire, see *America as Empire*, by Jim Garrison.

Hand in hand with America being seen as vulnerable, the world became aware of its dominant power. What had been overlooked or tolerated during preceding administrations became Europe's big concern: America, no longer only a nation among nations but now an empire among nations. That hurt. Many people—probably most people—don't like the empire. All over the world, sales of American flags have increased dramatically—flags to be burned in protest. And, incredibly, George W. Bush managed to lose a global popularity contest to Saddam Hussein. But make no mistake about it. We have only just entered a long era of the American empire. And for those who say it is not America we don't like, it is George Bush, it will not matter who is president; America will dominate the twenty-first century, as Rome did the first century. The only challenger will be China, but not for many years.

MULTILATERAL EUROPE VERSUS UNILATERAL AMERICA

There was a joke in Brussels when George Robinson was departing NATO as secretary general: He asks God, "When might NATO's capabilities improve?"

"Maybe in the lifetime of the next secretary-general," God replied.

Cheered, Javier Solana, European Union foreign policy chief, put the next question to God: "When might Europe have an effective foreign policy?"

Pause. Finally God replies: "Not in my lifetime."

If we take that as God's judgment on Europe's foreign politics, it is a pretty serious problem for a wannabe federation that, at least on moral and humanistic grounds, wants to run the world, especially if you consider that the role of the world power is already taken. It has become fashionable to say that America is a unilateralist, that America just does what it wants to do when it wants to do it, without proper consultation with others, without—like Europeans—taking a multilateral approach and consulting with others.

Americans don't like doing business by committee. The Kosovo war showed how difficult it was for America and its European allies to fight a war together. With the necessary coordination of *all* the NATO countries at every turn, America learned a lesson in the Allied war in Kosovo. In the words of General Wesley Clark, the American supreme commander of NATO forces, "We never want to do this again."

One can easily hear President George W. Bush saying the same thing about going through the United Nation's Security Council on certain matters: "We never want to do this again."

And, as Robert Kagan says: "Those who cannot act unilaterally themselves naturally want to have a mechanism for controlling those who can.... For Europeans, the U.N. Security Council is a substitute for the power they lack."

Contrary to some European perceptions, Bill Clinton ended up no less a unilateralist than George Bush. The Clinton administration had come into office talking about America as "the indispensable nation," as Secretary of State Madeline Albright never tired of saying.

In October 2002, a majority of the Democrats in the U.S. Senate supported the resolution authorizing President Bush to go to war with Iraq, while the socialists in France, Germany, Italy, and even the United Kingdom looked on in amazement. Kagan points out that in fact George W. Bush got more votes in the U.S. Senate to go to war in Iraq than his daddy did to start the first Gulf War.

You may remember that Ronald Reagan was routinely caricaturized by Europeans as a headstrong cowboy. Americans are "cowboys," Europeans love to say. And there is a lot of truth to this, although as someone who grew up in the West—in Utah—and rode horses from childhood until I joined the U.S. Marine Corps, I can't say being a cowboy is all bad.

America's counterpart to the cowboy label is that, to quote Kagan again, "Europeans simply enjoy the 'free ride' they have gotten under the American security umbrella over the past six decades. Given America's willingness to spend so much money protecting them, Europeans would rather spend their own money on social welfare programs, long vacations, and shorter workweeks."

So, Americans are cowboys, and Europeans are freeloaders.

At least Ronald Reagan rode horses. George W. Bush does not. He drives around his Texas ranch in his Ford pickup.

Well, obviously these stereotypic caricatures will get us nowhere. We have to get beyond them and seek common ground.

EUROPE 2010, A SUMMARY

Six years ago the European Union leaders committed themselves to creating "the most competitive and dynamic knowledge-based economy in the world by 2010." The spirit was euphoric, but during the race to reach the goal, struggles and hurdles cluttered the path.

The Hurdles

1. **High taxes and big governments.** EU governments take 45 percent of their national income, China just 11 percent, most other parts of Asia 20 percent, America less than 30 percent of the GDP.

2. **Less innovation.** The EU spent 1.9 percent of its national income for research and development compared with America's 2.6 percent. America has twice as many scientists and engineers per million people as the EU. And the scientific brain drain from Europe to the U.S. continues.

3. **Slow productivity growth.** The EU's business-sector productivity growth was only 0.6 percent during each of the last five years. America's was four times greater, and China's was four times greater than America's.

4. **Restrictive labor laws.** In the run for global economic supremacy, the height of this hurdle is set by the European Union. The rhetoric of loosening the restrictions needs political leadership for action.

5. **Declining export market share, and rising protectionism.** No change in results without a change in values. In spring 2006, France released a list of 11 industrial sectors in which the French government claims a veto right over foreign mergers. The European Union's export market shares have fallen to 42 percent from 51 percent in 1980; in high-tech products, the EU's share has fallen to 29 percent from 36 percent.

The Lisbon pledge breaks forth every spring, even as Europe continues in decline versus the United States. The EU's per-capita is now 25 percent lower than that of the United States—that gap has been widening for 15 years. The Organisation for Economic Co-operation and Development (OECD) says that if present trends continue in 20 years, the average American citizen will be twice as rich as the average Frenchman or German.

Indeed, the decline is built in.

In the beginning of the EU, the emphasis was on *openness*, opening up the countries to each other for unfettered trade and commerce. We all know that it is economic freedom, diversity, and competition that make a country or—in this case—a region competitive. But in recent years there has been a shift from openness to *harmonization*. Harmonize corporate taxes to the highest level of the biggest EU countries, harmonize to the most noncompetitive labor regulation, harmonize to the most restrictive immigration policies. In short, to harmonize all the European Union countries so that they will experience economic decline together—in solidarity.

It is Europeans and European politics that are shaping Europe's future. When I am asked what the future of Europe will be, the answer is that it is totally up to what the Europeans decide to do. If asked what my outlook is, it is that, while giving only lip service to economic reform, Europe will continue to hold on to its social model, and continue to pile up more "harmonizing" regulations. In other words, as of now, Europe is choosing the path of what I call "Mutually Assured Decline."

MINDSET #3   Does "Mutually Assured Decline" sound too harsh an economic judgment for the EU? Not if you have applied Mindset #3, "Focus on the score of the game," as I have for the past seven years while living in Europe. There has been a huge collective disconnect between political rhetoric and economic performance. Beware that this disparity is true of almost all countries, but the EU and its members take first prize in inaction and duration. Until Europe gets some real leadership, it will continue on the M.A.D. path.

MINDSET #4   The other side of the Mindset "Understand how powerful it is not to have to be right" is defending to the death the rightness of your position. Politics is driven by having to be right and proving the other party wrong. Although not limited to Europe, the more government-controlled an economy is, the more it celebrates being right.

MINDSET #9   "You don't get results by solving problems but by exploiting opportunities": Simply said, European politics is full of problem solvers. Too many bureaucratic rules leave too little room for opportunity seekers to bring about real change. Worse, new European Union members like Slovakia and Slovenia, where entrepreneurship is invited by competitive low tax rates and supported by a nourishing business environment, are threatened by old members to be leveled down to the least competitive common denominator. But without active and rewarded entrepreneurship, an economy is destined for stagnation.

Chapter 5

OUR EVOLUTIONARY ERA

Reservoir of Innovation

We are the beneficiaries of the revolutionary breakthroughs of the last years of the twentieth century. The next half of this century will be an era of absorbing, extending, and perfecting those great breakthroughs. Like the elephant nourishes the boa throughout the digesting process, the inventions of the revolutionary period feed a spirit of innovation.

Hype for the Next Big Thing fills space in the media, but it does not inspire a single idea, certainly not a Next Big Thing, which won't come any time soon. And those who keep waiting for it will be missing out on the great opportunities right in front of their noses.

TIME TO TAKE THE PROFIT

Peter Drucker called the last years of the twentieth century a time of "discontinuous change." Signs of discontinuous change are changes not seemingly connected to each other, things coming from out of the blue, radically new concepts, surprises, and upheavals that in political, societal, and economic currents take years to adapt to. In the first years of the twenty-first century, the world has moved from a period of discontinuous change to a long period of continuous change. Today, mostly unrecognized, mostly underestimated, we are deeply in the middle of what I call an evolutionary era, a period that builds on a ground already prepared.

In such a stage of history, waiting for the Next Big Thing is like waiting for something to fall out of the sky. It is like waiting for Godot: Godot never comes. If we continue to wait for Godot, our view of the matrix of history will fade. It is a mindset oriented to the past, but without history's lessons.

What does history tell us? I will take you on a short journey through next big things—shifts that had revolutionary impact, followed by very long evolutionary periods.

A HISTORY OF NEXT BIG THINGS

If the history of inventions and major innovations were a seismographic printout, we would see that at certain points the recorder swings out profoundly for a while, then, for a much longer period, constant but modest movements are recorded. The profound movements occur of course when there are major revolutionary breakthroughs, shaking the ground. What follows are long modest movements in which the churned-up ground settles again. These are the evolutionary periods during which the world profits in a greater way than during the initial revolution. In the time frame of the past seven centuries, revolutions have come in clusters, almost as if the universe had spilled out a spirit of rebirth.

The time in which the cornucopia of intellectual spirit spilled out most plentifully was the Renaissance—the great liberating revolution of the mind after most of the earlier advancements of the Middle Ages had virtually ceased during the Middle Ages. Starting in northern Italy, the Renaissance spread its liberating spirit through the Europe of the fourteenth through the sixteenth centuries, introducing printing and starting great universities, including Oxford, Heidelberg, and Vienna. In the section on the Mindset "Don't get so far ahead of the parade that people don't know you're in it," I wrote that Copernicus's and Galileo's ground-breaking insights were still too far ahead, even in the climate of the Renaissance that later gave us Leonardo da Vinci, Shakespeare ... and the postal service.

The Renaissance changed our minds and opened them to great new possibilities, all of which took a while to settle down until the seismic pointer was dramatically swinging out again. To get to the next big outswing, we have to make a big step to the end of the eighteenth century.

If we call the Renaissance the great liberation of the mind, we could call the next revolutionary period the time of inventing beneficial technologies, probably technologically the biggest "next big thing" time in history. It was the Industrial Revolution, which began in the last years of the eighteenth century. Led by the invention of the steam engine, the Industrial Revolution quickly got us started on the road to factory-filled cities, gas street lighting, steam ships, railroads, the first unions, free trade, baseball, department stores, photography, typewriters, sewing machines, the Suez Canal, and the first university degrees granted to women in America (1841). After the ground had been shaken by these revolutionary inventions and innovations, a long period of absorbing, extending, and perfecting followed. We have been absorbing, perfecting, and profiting from the products of the Industrial Revolution ever since.

At the end of the nineteenth century, the spirit of rebirth again was in the air. The next cluster of revolutionary advances appeared. During a short period of two decades, in the 1880s and 1890s, five amazing discoveries were made that in time led to the modernity we know today: electricity, the telephone, the automobile, aviation, and radio.

- Automobiles and airplanes became the new platforms for transportation.
- Electricity became the new energy platform for running almost anything.
- The telephone and radio became the new platforms for communication.

More than a hundred years later, we are still building on these breakthroughs; the evolutionary process of altering and perfecting has not come to a halt. The real profit of a revolution is often earned not by the inventor but by those who perfect and extend the invention.

On a big scale, China—for more than a millennium, until the middle of the fifteenth century—could be called a country of revolutionary breakthroughs. The list of inventions includes the compass, gunpowder, paper and printing, oceangoing ships, the rigid horse collar, porcelain, and the water clock (although there is no consensus where the first was made—in Mesopotamia, India, or China). China did not take the opportunity to evolve any of those breakthroughs to greater potential. I want to pause and take the clock as an example both of China and of the long time line that evolution takes from the beginning idea to its full development and use. And we never know when the full development of an idea will have been completed.

THINGS WE EXPECT TO HAPPEN ALWAYS HAPPEN
MORE SLOWLY

In searching for an example of how long an evolutionary process can last, I stumbled across Chinese water clocks. Until that time, I had not even known that there had been such clocks. I found the first records of modest timekeeping going back to the moving shadow of Egyptian obelisks built as early as 3,500 BC, forming a primitive sundial. Two thousand years later, the Egyptian sundials were able to divide a sunlight day into 10 hours and 2 twilight hours.

Early records of water clocks go back to 1500 BC, some even further. More elaborate mechanized astrological clocks were developed by the Greeks, who called them clepsydras (water thieves), by the Romans, and in the Far East from 100 BC to AD 1300. All these water clocks drove various mechanisms that illustrated astronomical phenomena. Some rang bells and gongs, others opened doors showing people or moved pointers, dials, and even astrological models of the universe.

During the Shang Dynasty, 1600 to 1066 BC (authorities differ on the dates), the Chinese water clocks were working in a quite sophisticated way. But the Chinese kept the secret of how they worked for royal astrologists, as the clocks were seen as exclusively for sovereignty and not for commoners. In 1088 BC, Su Sung (蘇頌) and his associates had built one of the most elaborate clock towers. There is a legend that a mistake created this clock. Su Sung, its constructor, was one of the emperor's advisors. He had to carry a birthday greeting to an emperor in another country. Su Song arrived one day early. That proved the Chinese calendar to be inaccurate. Su Song was so embarrassed that he asked his emperor to let him make a better clock to keep track of dates and times. It took Su Song seven years to built a wonderful water clock tower. It was three stories high and over 35 feet tall, with an armillary sphere on the roof and a celestial globe on the third floor. It possessed a bronze, water-driven armillary sphere for observations, an automatically rotating celestial globe, and five front panels with doors that permitted the viewing of changing manikins that rang bells or gongs and held tablets indicating the hour or other special times of the day.

In Europe the evolution of timekeeping began at the end of the thirteenth century, in the very beginning of the Renaissance. Through the first half of the fourteenth century, large weight-driven mechanical clocks, regulated by a verge-and-foliot escapement that controlled the speed of the clock, began to appear in the towers of several large Italian cities. Much like its Chinese predecessors, the Catholic Church would have loved to have a monopoly on timekeeping. But the evolution continued.

Between 1500 and 1510, Peter Henlein of Nuremberg invented springs to replace the heavy drive weights, permitting smaller portable clocks and watches. In 1656, Christiaan Huygens, a Dutch scientist, made the first pendulum clock, regulated by a mechanism with a "natural" period of oscillation. Around 1675, Huygens developed the balance wheel and spring assembly, still found in some of today's wristwatches. His improvement allowed portable seventeenth-century watches to keep time accurately to within 10 minutes a day.

The Web site of the National Institute of Physics and Technology continues the story:

> ...in London in 1671, William Clement began building clocks with the new "anchor" or "recoil" escapement, a substantial improvement over the verge because it interferes less with the motion of the pendulum.
>
> In 1721, George Graham improved the pendulum clock's accuracy to 1 second per day by compensating for changes in the pendulum's length due to temperature variations. John Harrison, a carpenter and self-taught clock-maker, refined Graham's temperature compensation techniques and developed new methods for reducing friction. By 1761, he had built a marine chronometer with a spring and balance wheel escapement that met the conditions of the British government's ... prize. [After many quarrels and difficulties, he finally received the 20,000 pounds in 1777] for a means of determining longitude to within one-half degree after a voyage to the West Indies. It kept time on board a rolling ship to about one-fifth of

a second a day, nearly as well as a pendulum clock could do on land, and 10 times better than required to win the prize.

... [Siegmund Riefler's refinements in 1889 led to a clock] with a nearly free pendulum, [and a] truly free-pendulum principle was introduced by R.J. Rudd about 1898.... One of the most famous, the W.H. Shortt clock, was demonstrated in 1921 [and] almost immediately replaced Riefler's clock as a supreme timekeeper in many observatories....

The performance of the Shortt clock was overtaken as quartz crystal oscillators and clocks [developed in the 1920s.] ...

Such quartz clocks and watches continue to dominate the market in numbers because their performance is excellent for their price. But the timekeeping performance of quartz clocks has been substantially surpassed by atomic clocks....

The first practical cesium atomic frequency standard was built at the National Physical Laboratory in England in 1955,... in collaboration with the U.S. Naval Observatory.... By 1960, cesium standards had been refined enough to be incorporated into the official timekeeping system of the [National Institute of Standards and Technology (NIST)].... As of January, 2002, NIST's latest primary cesium standard was capable of keeping time to about 30 billionths of a second per year....

This short trip through the history of how we measure time might disillusion those who expect speedy future evolutionary processes. It is at least a caution not to be too impatient. Much of the modern world would not be possible had the evolutionary work in timekeeping not taken place. But most who had a share in the evolutionary path of the clock didn't make money on the way. Many steps in any field of evolution cause widespread excitement and enthusiasm—and burst a lot of bubbles.

TIME TO TAKE OFF

If we take a look at the evolution of new transportation platforms, we see that many sought their chance. In the United States, in the early years of the automobile as we were working to figure out fuel, size, design, and replacements, the automobile industry was building up, and more than 2,700 car companies were created. During a long shakeout period, almost all of them went bust. Today in the United States, there are only 4 automobile companies, and only 2 of those are American companies. More than 99 percent of the 2,700 companies did not make it.

The evolution of the other transportation platform—aviation— was helped a great deal by the military. Quite apart from the political aspects, we should note that war and the military have over the centuries played a big role in change and in shaping the future. War has always been a great theater for perfecting innovations. It seems to be part of the human condition that when we are at our most destructive, we are sometimes at our most inventive. In the present case, it is the United States that leads the innovation parade.

America introduced digital warfare during its fights in Afghanistan and Iraq, including the use of unmanned aircraft. The Pentagon predicts that in just 20 years, 90 percent of combat aircraft will not have pilots on board. In Afghanistan, the United States for the first time flew armed aircraft into combat with no pilots in them. They were controlled by pilots in the United States, 10 time zones away. It will probably take some time to get the public to accept pilotless commercial airline travel. But the idea of pilotless transoceanic cargo flights may be in place by 2010.

There is a joke in aviation circles that in the future there will be only a pilot and a dog in the cockpit. The pilot's job is to feed the dog. The dog's job is to bite the pilot—if he touches anything.

The development of the heavier-than-air transportation platform has been in an evolutionary process for 100 years. And during those 100 years, accumulatively, not a penny has been made in the airline industry since the Wright brothers' flight. Aviation is the world's oldest bubble. No wonder investment guru Warren Buffett said that had he been at Kitty Hawk, he would have felt it his responsibility to shoot the Wright brothers down on their first flight. When he recently did invest in an airline for a short period, he pleaded "temporary insanity."

Part of the reason no money has been made in aviation is that each country had to have its own airline—a flag, a currency, and an airline. As the perfection of the invention and the waiting for airlines to make money continues, many questions in building aircraft are still open. In 2003, I was involved with the capacity issue.

Boeing and Airbus have been locked in a titanic struggle for supremacy over many years. What is the future of traveling?

To Airbus it is the A380, with a capacity of up to 800 passengers to build on the traditional airport hub system that would deliver more passengers to the hubs of their customers, the airlines.

To Boeing, it is the 787, a *smaller* plane that holds fewer passengers to—in an ever-shrinking world—get travelers faster and more directly to and from destinations.

My bet is on the side of Boeing, which says, "Come with us on a smaller, faster plane flying directly to your destination." Surely the last thing travelers want is to be jammed in with even more people traveling to a hub. But the market will decide.

All these considerations are made not by those who invented the technology but by those who are evolving it. Airbus did not design the first wide-bodied passenger jet, Boeing did not invent the jet engine, Dell did not invent the personal computer, Campbell's did not pioneer the concept of canned soup, and McDonald's did not invent the hamburger or the fast-food restaurant. All have very successfully evolved the original concepts.

TIME TO DIGEST

After the direction-setting breakthroughs of the Industrial Revolution, it took the world 100 years to digest what had been created in the last years of the nineteenth century. Much of the first half of the twentieth century passed without a new next big thing; it only started in the last years of the century. It was a period of evolution, with everyone making their products better or different. Products developed in the interest of military and government use were improved significantly. But even huge incremental changes to existing products or services are not next big things. The Internet boom was not a period of explosive invention, of next big things; it built on a technology already created.

- Television began just before the middle of the twentieth century, but it was seen as an extension of radio. We did not understand until much later that it would have such an impact on our lives and societies.
- The Soviets put up *Sputnik* in 1957, which we thought was the beginning of the space age, but it was really the beginning of the globalization of the information age through communication satellites.
- The 1940s saw the first controlled, self-sustaining fission reaction produced under Stagg Field at the University of Chicago by a team led by physicist Enrico Fermi, leading to the development of the atomic bomb and nuclear energy.

Many of us have lived through the last "cluster of inventions," and some even had a share. It took place in the last two decades of the twentieth century, especially in the 1980s and 1990s. We saw signs of discontinuous change in political, social, and economic life: information technology, biotechnology, nanotechnology, the new China, the new American empire, the new Europe, globalization, and privatization. The 1980s and 1990s were two great decades of revolutionary advancement.

Compared with the time it took from the primitive mechanical water clocks to the atomic clocks, the evolutions that have taken place since the Renaissance have speeded up a great deal.

Engineer Charles Parsons was responsible for a great breakthrough invention, the turbine generator, in 1888. For the next 40 years, his company followed a pattern of steady incremental innovation through a series of minor design and engineering improvements. These were not dramatic, but they were sufficient to give CA Parsons & Co. four decades of worldwide leadership in its core market.

NOT EVEN THE RAZOR WAS BUILT IN A DAY

It took 70 years from the invention of the safety razor by King Gillette in 1901 to the launch of his company's two-blade version in the 1970s. Since then, the evolution of the safety razor at Gillette has been a turned-backward history of onion peeling. One blade added to another.

Dan Roberts, the U.S. business editor of the *Financial Times*, recently wrote an amusing column tracking that history. Three blades, he tells us, became the world's top sellers when launched as the Mach3 Turbo in 1998. Gillette's big rival, Wilkinson Sword, went to four blades with the Quattro in 2003. And sure enough, Gillette jumped to five blades with the Fusion in 2005. Roberts reports that the obvious invitation to satire was taken up by the *Onion* newspaper when it ran a pretend article by James Kilts, Gillette's CEO, suggesting this was the only way to get the company's—as it were—edge back: "What part of this don't you understand? If two blades is good and three blades is better, obviously five blades would make us the best razor that ever existed. Comprende?"

As Roberts concludes, "the [five-blade] Fusion suggests we may be reaching the limits of genuine innovation in some product categories." It may be that the safety razor is now about ready for the next big thing, but evolution has served it pretty well for more than 100 years.

NANOTECHNOLOGY, THE NEXT BIG THING?

Manufactured products are made from atoms. The properties of those products depend on how those atoms are arranged. If we rearrange the atoms in coal, we can make diamonds. If we rearrange the atoms in sand (and add a few other trace elements), we can make computer chips. If we rearrange the atoms in dirt, water, and air we can make potatoes.

Nanotechnology researcher Ralph Merkle says that:

[T]oday's manufacturing methods are very crude at the molecular level. Casting, grinding, milling and even lithography move atoms in great thundering statistical herds. It's like trying to make things out of LEGO blocks with boxing gloves on your hands. Yes, you can push the LEGO blocks into great heaps and pile them up, but you can't really snap them together the way you'd like.

In the future, nanotechnology will let take us off the boxing gloves. We'll be able to snap together the fundamental building blocks of nature easily, inexpensively and in most of the ways permitted by the laws of physics. It will let us fabricate an entire new generation of products that are cleaner, stronger, lighter, and more precise.

Nanotechnology should lead to:

- Getting essentially every atom in the right place
- Making almost any structure consistent with the laws of physics that we can specify in molecular detail

- Manufacturing costs that do not greatly exceed the cost of the required raw materials and energy

Nanotechnology is just getting started, a speck on the horizon. Actually, it's less than a speck—the building blocks being invisible. We are no
further than in the Egyptian period of primitive sundials. But money is
pouring in and hopes are flying high. The results so far stay between
meager and nonexistent.

A few metal coatings have been created, and also some stain-
resistant clothing and some nanotubes. Cylindrical molecules of carbon
known as nanotubes are the strongest material known, and scientists
have now spun yards of thread made of almost 100 percent nanotubes.

Will we have a boom and bust like that of the dot-coms?

Yes, almost certainly—but not for some time.

In the meantime, look for more investment hype and apocalyptic
science fiction like Michael Crichton's book *Prey*, about nanotech run
amuck, or Hewlett-Packard's ad about how the company will be able to
make a cell phone small enough for an ant.

Do not be mistaken—even as the process of evolution speeds up,
many revolutionary breakthroughs are still in a pretty primitive stage.
Even the Internet. And that is good news. It offers a lot of perfecting and
extending to entrepreneurs—and risk, as we know. Given the nature and
character of information technology, we will probably end up with thousands of companies after we have gone through a long shakeout period,
but we will have to go through hundreds of thousands of companies to
get there. The revolutionary gifts of the twentieth century have set the
agenda for the innovations in the twenty-first century.

It is unlikely that there will be a revolutionary next big thing any
time soon, although with the acceleration of change, the next cluster of
inventions could be in only 75 years, or even 50 years, but certainly not
just ahead of us. Evolutionary trends have crowded the pipeline and will

keep us occupied during the first half of this new century, extending and perfecting them. It is like the elephant in the boa constrictor. The boa swallows the elephant, and we can see the big lump in the boa as it moves through the boa's body, just as the big lump of revolutions are slowly digested and perfected, getting smaller and less nourishing as they slowly move, until eventually the next elephant comes along.

TIME TO FACE THE DARK SIDE ...

In response to revolutions, the evolutionary process can produce side effects of unintended consequences, the boomerangs in the ecology of technology. Before the invention of the telephone, the standard way to communicate was through letters. That's why in Dublin, Ireland, at the end of the nineteenth century, there were eight mail deliveries a day. I remember—in my younger, loftier days as a graduate student—reading the letters of poet William Yeats and Lady Gregory. In passing, I noticed that the letters were sent, responded to, and responded to again, all with the same day's postmark. It was not until years later that I learned of all these mail deliveries. With the advent of the telephone, the number of mail deliveries went down until—eventually—there was only one each day. The consequence, the unintended one, was the death of the art of letter writing.

... AND THE GOOD SIDE

Besides the regrettable decline of the written word as an unintended consequence of technology, which of course is also supporting our rush into a visual world, there is an upside to the cost of communication as well.

In England up to the 1840s, if you wanted to send a letter you had to go to a post office where an army of clerks figured out what to charge you—depending on the distance the letter was to travel. Then along came a genius named Rowland Hill, who figured out that all the work was done at the sending and receiving post offices, and it didn't matter how far the letter traveled. He said that each person should be charged the same price for sending a letter. And because it was the same price, that price could be paid ahead of time. Thus was the penny post born.

Today, the same is true of the telephone companies. It cost them the same to send a call across the street or around the world. But they continue to charge us by the distance traveled. Prices are coming down dramatically because of competition, especially calls on the Internet. The world in turn will be dominated by person-to-person communications. And all calls will be local.

Most of the speculation in the past 30 years has focused on what I call the hard technologies: computers, a wired and wireless world, telecommunications—technologies that are around us in our homes and workplaces and change the way we do things in an often dramatic way. But the most consequential technologies of the twenty-first century will be the "soft" technologies with an impact not limited to how we live and work but questioning what we are. For millennia, the ultimate authority of our meaning and destination was God. With Darwin, the origin of mankind had come into question. With the new biotechnologies, we have reached the turning point of creating our own destiny.

TIME TO RAISE THE QUESTION

Germline engineering—where changes are made in the genetic codes that are passed on generation to generation—will overwhelm the importance of all previous technologies. But it will also include the danger of catapulting the human race into an undreamed future.

The great dilemma of the twenty-first century will be that although germline engineering will allow us to treat and eventually eliminate diseases and disorders such as Alzheimer's, Down syndrome, and Parkinson's, the very same technology will allow us to make people taller, stronger, smarter, more beautiful. In short, we will be able to create a perfect race. This of course falls under the long shadow of eugenics, the perfection of the human race. Hitler had the idea, but he did not have the science. Here comes the science!

Once the first step is made, we will be on a path of no return. Dispute over this matter will lead to a huge confrontation between science and religion, between feasibility and humanity. It is a confrontation shaking up basic beliefs and values such as during the times of Galileo and Darwin.

With the mapping of the human genome, the instruction book with the codes for how the physical human is put together and how that can be changed, scientists will have the key to our physical makeup. The mystery will be gone. But I think we should not be too sure. What scientists will not have is the key to our souls, our spiritual nature, which we will therefore cling to and obsess about. The great question of this new century is raised by the spirit of the past, although with a new urgency: What does it mean to be human?

Darwin, deeply convinced of the transmutation of species, was well aware of the blasphemous character his insights had for a devout world. The Christian was struggling to obey God; the scientist was fighting for the laws of nature.

Few will face such a deep controversy of mind and soul, scientific proofs and religious beliefs. But we all are challenged to find a balance between progress and constancy, between ambition and meditation, between the profusion of high tech and the lack of high touch, what we make of our lives. We take care of what we physically possess, but we often forget to nourish our souls.

In one of our conversations in Cambridge, where I lived for many years, Peter Senge introduced me to a letter written by the man who shook the foundation of our origins and destiny, the letter Charles Darwin wrote at the end of his life:

> Up to the age of 30, or beyond it, poetry gave me great pleasure. But now for many years I cannot endure to read a line of poetry. My mind seems to have become a kind of machine for grinding general laws out of large collections of facts, and if I had to live my life again, I would have made a rule to read some poetry and listen to some music several times every week. The loss of these tastes is a loss of happiness, and may possibly be injurious to the intellect and more probably to the moral character, by enfeebling the emotional part of our nature.

Whatever the future holds, it will be worth nothing if there is no joy.

APPLYING MINDSETS

MINDSET #11   "Don't forget the ecology of technology": If necessity is the mother of invention, change is the father of innovation. There is a reason why "growth through innovation" is the new mantra of business. The evolutionary period we live in is the time to "pick the ripe fruits" that were seeded during the revolutionary period in the last years of the twentieth century. To "Consider the ecology of technology" also means to open your mind for all the opportunities that are embedded in the slow digestion of the last Big Things.

MINDSET #4   "Understand how powerful it is not to have to be right": Nothing ventured, nothing gained. If you fear to be wrong, you will not be able to exploit the opportunities this evolutionary period is offering.

Remembering Einstein, I would compare this era to an orchard where a great variety of fruit is ripening. While problem solvers are still worrying about the fruits that have already fallen, opportunity seekers look for the new ones that are ripe for the picking. This reservoir of innovation built on the revolutionary period at the end of the last century will supply us with fruit for most of this century.

ENDNOTES

Mindset #1: While many things change, most things remain constant

4 Quote from Will and Ariel Durant, from their series of books *The Story of Civilization*, New York: MFJ Books, 1993.

6 Nat Holman quote and Hank Luisetti story drawn from the Luisetti obituary in the *New York Times*, December 22, 2002.

8 Jim Collins and Jerry Porras quote and story from *Financial Times*, August 5, 2005.

8 King Solomon quote from Ecclesiastes 1:9.

Mindset #2: The future is embedded in the present

11 Excerpt of "The Story of Johnny Look-in-the-Air" from Heinrich Hoffmann, translated by Mark Twain.

17 The 7.8 percent figure is from the U.S. Bureau of Labor Statistics, published in the *New York Times*, January 29, 2006.

Mindset #3: Focus on the score of the game

23 Excerpt on General Motors from *Business Week,* May 9, 2005.

24 Bill Ford's statement was made on *CBN NewsWatch*, January 27, 2006.

25 Story and numbers on the Nobel Prize from "Business Can Be a Nobel Pursuit," *Financial Times*, October 10, 2002. Nobel Prize winners for the years 2003, 2004, and 2005 added [13 for the United States and 5 for Europe].

25 Quotes on global cooling from Symposium on Sustainable Development, Vienna, Austria, February 22, 2002, background papers, sponsored by the Austrian government.

26 British airways story from the *Economist*, October 29, 2005.

27 Polar bears story from Jim Carlton, *Wall Street Journal*, December 14, 2005.

28 Story on the dying species from the *Seattle Times*, September 12, 1999.

28 Story on Dr. Myers's estimates from the *New York Times*, August 19, 2001.

29 George Burgess quote from the International Science News Web site, unisci.com.

30 Coconut story taken from http://www.cbsnews.com/stories, May 21, 2002.

31 Chernobyl figures and story from George Melloan's column in the *Wall Street Journal Europe*, October 11, 2005.

Mindset #4: Understanding how powerful it is not to have to be right

53 Quotes about Albert Einstein from "Das Wunder von Bern," *Der Spiegel*, March 2005.

33–36 Letters from Albert Einstein from *The Collected Papers of Albert Einstein. Volume 5. The Swiss Years: Correspondence 1902–1914*, Princeton University Press, Princeton, NJ, 1993.

38 Dr. Barry Marshall's quotes and story from "A Nobel Prize for Intuition," by Madeline Drexler, *International Herald Tribune*, October 8, 2005.

Mindset #5: See the future as a picture puzzle

42　Salman Rushdie quote from an article in the *Wall Street Journal Europe*, September 27, 2005.

42　Sir Isaac Newton quote from Henry Pemberton, "A view of Sir Isaak Newtons Philosophy," London, 1728.

43　Dr. Jürgen Renn quote from the Max-Planck Institute online: http://www.mpiwg-berlin.mpg.de/. Last accessed 2005.

Mindset #6: Don't get so far ahead of the parade that people don't know you're in it

48　Poem writtten by Johann Christian Fürchtegott Gellert, 1715–1769.

48　Eratosthenes assumed the sun to be far enough away that its rays were parallel, so he measured shadows in two cities at noon. He knew the distance between the cities. With this information, he could calculate the circumference of the Earth as 250,000 stadia (a Greek measure that varies from 154 to 215 meters). His estimate comes as close to the real value as possible.

50　Johannes Gutenberg's Bible, published September 30, 1452, became the first book to be published in volume.

50　Darwin's work during the *Beagle* expedition allowed him to study first-hand geology, fossils, and a multitude of living organisms, as well as meet native people. He methodically collected an enormous number of specimens, many new to science, which established his reputation as a naturalist and made his work a precursor of ecology. His detailed notes formed the basis for his later work as well as providing social, political, and anthropological insights into areas he visited.

Mindset #7: Resistance to change falls if benefits are real

57　*"Liberté, Egalité…"* means Freedom, Equality, Brotherhood, or death! Regulation, Equalization, Bureaucracy, that is death!

57　Valéry Giscard d'Estaing quote from *Le Monde*, April 20, 2005. Former

French president d'Estaing chaired the drafting of the European Union constitution.

Mindset #8: Things that we expect to happen always happen more slowly

65 Information on fixed wings from Toni Ofner, who holds a patent on an alternative wing system.

66 Arthur C. Clarke's predictions from *The Futurist*, March–April 2000.

67 Numbers on the spread of the telephone from http://www.triotel.net.

68 Story on the Lumière brothers and Louis Lumière quote from "Confounding Machines: How the Future Looked," *New York Times*, August 28, 2005.

68–69 Story on the economic development of Lumière brothers enterprise from http://www.imagesjournal.com/issue04 and http://www.cinescene.com. Last accessed 2006.

69 D.W. Griffith quote from "Confounding Machines: How the Future Looked," *New York Times*, August 28, 2005.

70 Information about the history of radio from http://earlyhistoryofradio.us/index.html.

71 J.M. McKibben quote from "Confounding Machines: How the Future Looked," *New York Times*, August 28, 2005.

71 Waldemar Kaempffert quote from "Confounding Machines: How the Future Looked," *New York Times*, August 28, 2005.

72 David Sarnoff quote from "Confounding Machines: How the Future Looked," *New York Times*, August 28, 2005.

72 Thomas Hutchinson quote from "Confounding Machines: How the Future Looked," *New York Times*, August 28, 2005.

72 The history of television was taken from http://www.inventors.about.com/library. Last accessed 2006.

73 T.S. Eliot quote from "Confounding Machines: How the Future Looked," *New York Times*, August 28, 2005.

Mindset #9: You don't get results by solving problems but by exploiting opportunities

77 George Bernard Shaw quote from www.shawalphabet.com.

78 Schwarzenegger quote is from CNN. It can be found at www. transcripts.cnn.com.

79 Churchill's statement about an iron curtain was made to President Truman in a telegram and later in a speech in Missouri in 1945.

81 Kohl quote from http://www.berliner-lesezeichen.de/lesezei/Blz97_01/ text26.htm.

83 T. Gary Rogers's story, "A Rocky Road to Vanilla," from the *New York Times*, May 2, 2004.

83 Jean-Jacques Doeblin quote and story from "Danone Wins by Being First Past the Poles," by Jan Cienski, *Financial Times*, May 31, 2004.

85 Rich Karlgaard quote from *Forbes*, February 9, 2004.

91 Maxine Martens's company is Martens & Head, http://www .maxinemartens.com.

Mindset #11: Don't forget the ecology of technology

107 Number on Gore-Tex sales is from the Gore-Tex Web site: http://www. gore-tex.com.

Chapter 1: Culture: A Visual Culture Is Taking Over the World

114 The Song of King Solomon 7, 6–13, 1020 BC.

115 Sir Vidia Naipaul and Tunku Varadarajan quotes from the *Wall Street Journal Europe*, October 27, 2004.

116 The 2004 National Book Award fiction winner was Lily Tuck for *The News From Paraguay*.

116 Larry Kirschbaum quote and story from the *New York Times*, October 17, 2004.

116 Publishing titles figures from "Book Volume Down," *USA Today*, June 1, 2006.

117 The NEA report, including the quotes from Dana Gioia, can be found at www.nea.com.

118 Janet L. Robinson quote from www.nytco.com/investors_nyt_circulation.html.

119–120 Rupert Murdoch quote from the *Financial Times*, November 25, 2005.

122 Oliviero Toscani quote from www.olivierotoscani.it.

123 PQ figures from the *Financial Times*, May 18, 2005.

123 David Brenner quote and story from "Virtual Ads a New Reality on TV Shows," by Sam Lubell, *International Herald Tribune*, January 2, 2006.

124 Robert Hayes and Ben Evans quotes from "Design Walks the Walk," *Financial Times*, September 13, 2005.

125 Scott Morrison quote from the *Financial Times*, September 13, 2005.

125 Swatch story and quotes from http://www.swatchgroup.com.

126 Dylan Jones quote from "The Beauty at Apple's Core," *Financial Times*, September 13, 2005.

127 Steve Jobs quote and story from "Apple Cube Becomes a Modern-Day Relic," by Phil Pallon, *New York Times*, August 16, 2001.

127 Frank Nuovo quote and story from an interview in the *Financial Times*, September 13, 2005.

129 Kurt Foster quote from *International Herald Tribune*, February 16, 2005.

132–133 Quotes on the design hotel all from *Vanity Fair*, May 2000.

133 "The New Luxury" editorial from *Financial Times*, June 12, 2004.

134 Alex Coles quote from *Financial Times Magazine*, December 2005.

134 Information on Karl Lagerfeld designing iPod from *Wired*, January 5, 2006.

134 LeCave and Karl Lagerfeld quotes and basic information from "Come Into My World," *Financial Times Magazine*, September 2005.

136 Quote on Luis Vuitton from http://www.designboom.com.

136 Yves Carcelle quote and coverage of event from "Can Art Ever Sell Handbags?" *Financial Times*, December 2005.

136–137   Vanessa Beecroft and Peter Marino quotes from *Financial Times Magazine*, December 2005.

137   The Prada SoHo store was wrecked by a fire in March 2006.

137   Prada quotes and reporting from "Prada Fashions a Cutting Edge," *Financial Times*, September 13, 2005.

138   Frank Gehry quote from "Gehry Downtown," by Joyce Caruso, http://www.artnet.com/magazineus/frontpage.asp, August 7, 2001.

139   Kate Arthur quotes from the *New York Times*, February 20, 2005.

139   Dave Goldberg quote from "I Screen, You Screen: The New Age of the Video," *New York Times,* July 31, 2005.

140   James Paul Gee quote from the *New York Times*, December 4, 2005.

140   Eric Zimmerman quote and reporting from John Leland's "The Gamer as Artiste," *New York Times*, December 4, 2005.

141   Steven Spielberg quote from the *New York Times*, December 4, 2005.

141   Todd Wagner quote from "Hollywood Is Facing a Journey into the Unknown," by Joshua Chaffin, *Financial Times*, August 10, 2005.

142   Rick Finkelstein quote from "Hollywood Is Facing a Journey into the Unknown," by Joshua Chaffin, *Financial Times*, August 10, 2005.

143   Lisette Model quote from the Web site of the New York Institute of Photography: http://www.nyip.com.

146   Emma Crichton-Miller quote from "How to spend it," *Financial Times*, October 2004.

147   Auction prices from Souren Melikian's article in *Art+Auction*, October 2005.

147–148   Jacques Lipchitz was born in Lithuania in 1891 and moved to Paris at age 18 to study sculpture. Just before the outbreak of World War I, he became involved with the Parisian avant-garde and was among the first to apply cubist principles to the creation of three-dimensional forms. In 1940, he set up permanent residence in the United States, establishing a studio in Hastings-on-Hudson, New York. While his European work is distinguished by its focus on subjects ranging from acrobats, harlequins, and nudes to still-life objects, Lipchitz turned his attention to mythic and biblical themes after his arrival in the United States, and the scale of his works became increasingly monumental. He died in 1973. Today his sculpture pieces sell in the million-dollar range.

149 The *Art+Auction* quote and other information on the biennial are from the magazine's May 2005 issue.

149 Information on biennials is from "That's Show Business," *Financial Times*, October 2004.

150 Information regarding the Museum of Modern Art shows was taken from "Corporate Power," by Barbara Pollack, *Art+Auction*, December 2005.

150 Marc Porter quote is from *Art+Auction*, December 2005.

150 Friedhelm Hutte quote is from *Art+Auction*, December 2005.

151 Michael Klein quote from is from art critic Regina Hackett, *Seattle Post-Intelligencer*, May 24, 2004.

152 Leatrice Eiseman and Karen Beckwith quotes are from "The Revolution Will Be Colorized," *New York Times*, March 13, 2005.

153 Shi Zongyuan and *People's Daily* quotes and coverage are from "Fear of a 'Color Revolution' Behind China's Renewed Clampdown on Media Freedom," *Financial Times*, November 18, 2005.

154 Quote from Martin Pigott and reporting are from "Clearer Views of the Data Mountains," *Financial Times*, August 17, 2005.

Chapter 2: Economics: From Nation-States to Economic Domains

168 Bill Weldon quote is from "Big Numbers Destroy All Intelligence," *Financial Times*, September 27, 2004.

169 Life Optics: http://www.lifeoptics.com.

170 Frequent-flyer miles value from the *Economist,* January 8, 2005.

171 Quote on the Trabant is from http://www.team.net/www/ktud/trabi. html.

174 Information on San Antonio Spurs team composition is from *USA Today*, November 2, 2005.

175 Stan Davis introduced us to the concept of mass customization about 20 years ago—a stunning insight in 1987. For an excellent new summary, see his tenth-anniversary edition of *Future Perfect*, published by Addison-Wesley in 1996.

176 Howard Springer quote from the *New York Times*, March 13, 2005.

176 Information on McDonald's outsourcing from Missouri to Colorado is from "Welcome to McDonald's, May WE Take Your Order?" *USA Today*, June 18, 2004.

Chapter 3: China: The Periphery Is the Center

183–184   Deng's quote and story are from the *New York Times*, November 20, 2005.

185 Mao Tse-tung quote is from his speech on November 6, 1957, at the Meeting of the Supreme Soviet of the USSR in Celebration of the 40th Anniversary of the Great October Socialist Revolution.

188 Numbers on Chinese apparel production are from "China gegen den Rest der Welt," *Die Zeit*, May 4, 2005.

188–189   Information on when China will overtake United States is from the following estimate: GDP/C US = 40,100; GDP/C China = 5,600; growth rates: 3 and 7, which translates to 31.5 years after calculation.

189 Ben Stein quote is from the *New York Times*, July 31, 2005.

189–190   Numbers on Chinese growth are from "What's to Stop India and China?" *Economist*, October 29, 2005.

190 Hong Liang quote and information are from "China Economy Even Bigger Than Thought," *International Herald Tribune*, December 21, 2005.

190 Numbers, story, and quote on Alibaba are from "Standing up to Ebay," *Forbes*, April 18, 2005.

190 Analysys International published numbers on Alibaba in a report called "Quarterly Tracker on China's C2C Market 2005-Q3."

191 Numbers on China's cities are from "New Boom Towns Change Path of China's Growth," *New York Times*, July 28, 2004.

192 Reporting on the urban renaissance in China is based on information in the *New York Times*, November 25, 2005.

194–195 The translation of "ke kou ke le" is from the Web site of the American Chamber of Commerce, in a report from Baker & McKenzie.

195 Bob Shi quote and story are from "To Take China, Pepsi Turns Farmer," *Wall Street Journal*, December 19, 2005.

196–197 Giorgio Armani quote and story are from "Giorgio Armani to Add 2 Shanghai Stores in China," *Bloomberg Online*, April 19, 2004.

198 Richemont and Swatch reports are from *Financial Times*, November 13, 2004.

199 Shen Zhiwen quote on Chinese models from "China's New Models Take Off on Catwalk," *China Daily*, November 30, 2004.

200 Yu Zida quote from "China Design," *Business Week*, November 21, 2005.

201 Stefan Fritschi quote and reporting from "China Design," *Business Week*, November 21, 2005.

201–202 Richard Branson quote and reporting are from "Today Shanghai, Tomorrow the Universe," *Financial Times*, December 12, 2005.

202 Number of Chinese billionaires are from the *Financial Times*, October 13, 2005.

205 Schumacher quote from the *Financial Times*, September 25, 2004.

205 Quote on Chinese Formula One is from the *Wall Street Journal Europe*, September 17, 2004.

206 Bob Lutz quote and prediction from "China Looms as the World's Next Leading Auto Exporter," *New York Times*, April 22, 2005.

206 Information on the hourly costs of autoworkers are from the *New York Times*, April 22, 2005.

206 J.M. Noh quote and story are from the *New York Times*, April 22, 2005.

207 The Li Chow quote and story are from "Hollywood Sees a Future in China," by David Barboza, *International Herald Tribune*, July 4, 2005.

207 Jonathan Landreth quote is from "Hollywood Sees a Future in China," by David Barboza, *International Herald Tribune*, July 4, 2005.

210 Figures on China's foreign direct investment are from the *Wall Street Journal Europe*, January 20, 2006.

210–211 India and China's growth figures and story are from Simon Long's survey of India and China in the *Economist*, March 5, 2005.

Chapter 4: Europe: Mutually Assured Decline

216 Quote from former Commissioner Anna Diamantopoulou is from the Web site of the European Union: http://europa.eu.int/.

216 Quote on variety of social models are from "Charlemagne—Choose your Poison," *Economist*, October 1, 2005.

218 Vladimir Spidla quote is from the *Financial Times*, July 10, 2005.

218–219 Quotes on the social model from Joshua Livestro are from the *Wall Street Journal Europe*, October 28, 2005.

221 The necessary fertility rate to maintain a population on a constant level (called "replacement fertility rate") is 2.1 births per woman, because demographers only consider an offspring for replacement if it reaches the age of 15.

221 Quote on demographic development is from "Half a Billion Americans?" *Economist*, August 22, 2005.

224 Quote on Hayek is from Thomas Sowell at Stanford University, www.hayekcenter.org/friedrichhayek/hayek.html.

226 Michael Ignatieff quote is from the *New York Times*, January 10, 2003.

227 *America as Empire* was published by Berrett-Koehler, San Francisco, 2004.

228 Robert Kagan, *Of Paradise and Power*. Knopf, New York, 2003.

230 Number on unemployment benefit within the EU is from Web site of the Organisation for Economic Co-operation and Development: http://www.oecd.org/dataoecd/37/53/2504598.pdf.

Additional dates on unemployment benefits from the Web site of Euromärsche: http://www.euromarches.org/deutsch/04/0319_7c.htm.

Chapter 5: Our Evolutionary Era: Reservoir of Innovation

236–239 Information on the history of timekeeping is from http://physics.nist.gov/GenInt/Time/revol.html.

243 Charles Parsons example from "Think Small for a Change," *Financial Times*, August 25, 2005.

243 Dan Roberts quote from his article "Is This Really the Best a Man Can Get?" *Financial Times*, September 16, 2005.

243 Story on razor blades from "Is This Really the Best a Man Can Get?" *Financial Times*, September 16, 2005.

244 Information about nanotechnology from Ralph Merkle, PhD: http://www.zyvex.com/nano.

247 Penny Post story drawn from the works of Arthur C. Clarke.

249 Charles Darwin's letter at the end of his life and information on his life can be found at http://pages.britishlibrary.net/charles.darwin.

ACKNOWLEDGMENTS

In many ways a book is like a child. Once born, it needs nourishment and support to develop and mature. And as with a child, the outcome is a mix of what is in its DNA and the influence of education and environment.

Looking back, the DNA of this book was developing over many years; most of its genes were sort of free-floating before the process of arranging them in two complementary strands—the two parts of the book—got started. But once the structure was found, it quickly grew.

I begin these acknowledgement pages with my wife, Doris Naisbitt. I published my last book in 1999, and if not for her, that would have been it; I would have never written another one. I met Doris as the German-language publisher for my 1995 book *Megatrends Asia*. At the beginning of this century, in the year 2000, we were married in Dürnstein, a 1200-year-old village on the banks of the Danube. It was a new beginning for me, a new life in Europe with a beautiful, intelligent wife who took my hand to begin a new journey. This book is part of that journey. We conceived and wrote it together. It is as much her book as my book. I am forever grateful to her for getting me started again, not only with this book, but with a new and beautiful life.

ACKNOWLEDGMENTS

As many contribute to the development of a child, many have added their advice and good spirits to this book.

I owe a huge debt to Jim Levine, my agent, the best in the business, who navigated us through the publishing thickets to HarperCollins, and was a great critic and warm supporter throughout the process of book proposal and finished manuscript. He is not only an agent, he watched over the book like a mentor, providing great support and stepping in whenever necessary.

To our HarperCollins team. First Joe Tessitore, the president who sealed our deal at the Frankfurt Book Fair. Our first editor, Herb Schaffner, was a great champion for this book and (brilliantly) annotated suggestions in the entire manuscript before leaving to head a publishing program in another house. Editorial director Marion Maneker, our editor since, could not have been more supportive and helpful in shepherding the book through to publication. Also deep gratitude to the other members of the team for great personal and professional support: Libby Jordan, George Bick, Angie Lee, Larry Hughes, Cecilia Molinari, Jamie Brickhouse, Alex Scordelis, and Felicia Sullivan.

A very special thanks to our good friend Spencer Johnson, author of *Who Moved My Cheese?* not the least for suggesting *Mind Set!* as the title of this book. More than that, Spencer, who knows a thing or two about successful books, has been a strong supporter from the beginning, providing outstanding advice to keep the focus on the benefit of the reader.

Steve Rhinesmith, former ambassador to the Soviet Union, is a great friend and supporter who has been involved with most of my books. He read several drafts of the manuscript, and his feedback was invaluable. Glenn Miyataki, longtime president of the Japan American Institute for Strategy in Honolulu, created a forum of Japanese and American CEOs in January 2005 where I first introduced my mindsets to the business community. Stan Davis read the manuscript at an early and critical stage, giving me sage advice and guidance.

Thanks also to my daughter, Claire Schwadron, who runs our U.S. office and kept things humming throughout the last intensive year, and to my son, David, and his wife, Shelly, who read and commented on an early draft. David, a colonel in the U.S. Air Force, is now in Afghanistan, commanding a Provincial Reconstruction Team in Jalalabad.

In the European league, a special thanks goes to Tatjana Halek-Schroeder, author and journalist, who plays with words like Mozart played with notes. She was our inspiration in reaching out to poems and metaphors, questioned the manuscript with the critical voice of a journalist, and supported it with her love for the written word. Her husband, Daniel Schroeder, often stepped in to rescue us when our computer turned against us.

Sven Atterhed in Sweden, a 30-year friend, read and commented on drafts at several stages. His warm support and critical advice, as always, are greatly appreciated.

Our Vienna friends Toni Ofner and Maria Lehrl read the entire manuscript, made many thoughtful suggestions that improved our efforts. In the Introduction I wrote of Toni's questions that led to the mindset idea. Maria, CEO of Life Optics, the company discussed in Mindset #9, added many practical comments about the challenges of operating an international company out of an EU member state. Werner Langthaler, the CFO of Intercell, Austria's leading biotech company, and former McKinsey partner, sharpened our focus in many discussions and gave us enthusiastic support. Harald Stolzenberg, just retired CEO of La Prairie, shared his international experiences with us, We spent great times with Petra Goette and Enrico Cucchiani at Lake Como, playing with ideas at an early stage; Maxine Martens gave us insights behind the fashion facades, and we thank her for our interesting discussions with fashion designers in Europe and the U.S.

Doris's daughter, Nora Kutschera, added her experience as a media consultant for Eastern Europe, providing insights into the sometimes adventurous business attitudes and practices of new members of the EU.

Last but not least, during the last three years in Vienna we have been favored with two very bright research assistants from the University of Vienna, Fabian Schroeder and Ferdinand Rauch. Ferdinand, who, at the age of 20, is working on his PhD on innovation, stepped in when Fabian left to continue his studies in Paris, and was our reliable and ambitious researcher. He and his brother also brought the spirit and the critical voice of the young generation into the game. These young men, fluent in several languages, certainly are part of the talent that in the global economy will go wherever it is well treated.

INDEX

Megatrends, JOHN NAISBITT'S first published book in 1982, was at the top of the *New York Times* bestseller list for more than two years and sold more than nine million copies. His experience reaches from executive positions in the corporate world to appointments by Kennedy and Johnson, and as an entrepreneur. He is a former visiting fellow at Harvard University, a former visiting professor at Moscow State University, and currently a faculty member at Nanjing University in China. Naisbitt serves as a Distinguished International Fellow, Institute of Strategic and International Studies (ISIS), in Malaysia. He has lived in three continents and travels the world as one of its most sought-after speakers. He holds fifteen honorary doctorates in the humanities, technology, and science.